HUMAN POPULATION AND THE ENVIRONMENTAL CRISIS

EDITED BY

BEN ZUCKERMAN

DAVID JEFFERSON

Jones and Bartlett Publishers
Sudbury, Massachusetts

Boston   London   Singapore

Editorial, Sales, and Customer Service Offices
Jones and Bartlett Publishers
One Exeter Plaza
Boston, MA 02116
1-800-832-0034
1-617-859-3900

Jones and Bartlett Publishers International
7 Melrose Terrace
London W6 7RL
England

Library of Congress Cataloging-in-Publication Data

Human population and the environmental crisis / edited by Ben
    Zuckerman, David Jefferson.
         p.   cm.
      Papers presented at a symposium held at the University of
    California, Los Angeles, on Oct. 29, 1993.
      Includes bibliographical references and index.
      ISBN 0-86720-966-6
        1. Population--Environmental aspects--Congresses.   2. Sustainable
    development--Congresses.   3. Biological diversity--Congresses.
    4. Population policy--Congresses.   I. Zuckerman, Ben
    II. Jefferson, David.
    HB849.415.H86   1995
    304.2'8--dc20   95-21148
                                                                                          CIP

Acquisitions Editor: Arthur C. Bartlett
Production Coordinator: Joan M. Flaherty
Manufacturing Buyer: Dana L. Cerrito
Editorial Production Service: Book 1
Typesetting: Seahorse Prepress/Book 1
Cover Design: Marshall Henrichs
Printing and Binding: Malloy Lithographing, Inc.

A contribution of the IGPP Center for the Study of Evolution and the Origin of Life (CSEOL),
University of California, Los Angeles

Printed in the United States of America
99 98 97 96   10 9 8 7 6 5 4 3 2 1

CONTENTS

CHAPTER 6   **POLITICS AND SOCIETY: POLITICAL CHALLENGES OF CONFRONTING POPULATION GROWTH   81**

Anthony C. Beilenson

CHAPTER 7   **GLOBAL ENVIRONMENTAL ENGINEERING: PROSPECTS AND PITFALLS   93**

Richard P. Turco

INDEX

PREFACE

The power that derives from science and technology has enabled human beings to dominate and drastically alter the biosphere. Unfortunately, the wisdom embodied in our cultural and political institutions has not kept pace, and the human race now faces worldwide problems of a type never before encountered.

Early in history, humans evolved the capability to deal with such short-term crises as the sudden appearance of stampeding bison or of an angry bear at the mouth of the cave-dweller's abode, but longer-term quandaries were largely irrelevant because their time scales were much longer than a human lifetime. Even the Ice Ages—certainly a major environmental crisis faced by early humans—occurred over tens of thousands of years, a period brief in terms of geological processes but unimaginably long in human terms. Yet over the past two centuries and especially in recent decades, the situation has changed markedly. The exponential growth of human population and technology has generated significant changes with which neither biological evolution nor our political systems can deal effectively. The following essay, "How Many Angels Can Dance on the Head of a Pin?" originally appeared in 1991 on the op-ed page of the *Los Angeles Times*. Written by an astronomer (and coeditor of this volume), it quantifies some relationships between human growth rates and both intermediate- and long-term events.

On October 29, 1993, a symposium addressing the relationship between "Human Population and the Environmental Crisis" was held on the campus of the University of California, Los Angeles. Convened by the UCLA Center for the Study of Evolution and the Origin of Life, the symposium was attended by diverse elements of the Los Angeles community including high school and college students, teachers and university faculty, researchers and concerned citizens, and members of the UCLA family and of the public at large. This volume makes accessible the papers presented at that symposium.

The expertise of the seven symposium speakers, each of whom contributed a chapter to this volume, spans the broad scope of the population/environmental problem. Dr. Mildred Mathias and Mr. Jean-Michel Cousteau address the biological and ecological aspects; Drs. Harte, Schneider, and Turco consider the physical, especially the atmospheric environmental, ramifications; and Ms. Jodi Jacobson and Congressman Anthony Beilenson are concerned primarily with major political and cultural questions. Because of the interdisciplinary nature of this all-encompassing subject, the seven chapters are necessarily interrelated. Each focuses on a definable aspect of the problem, and each emphasizes a particular perspective, but all share a common view: Population growth has had, and at present continues to have, a severe impact on our environment. The problem is real, immediate, and *demands* solution.

HOW MANY ANGELS CAN DANCE
ON THE HEAD OF A PIN?

How many angels can dance on the head of a pin? This is a type of question that people asked in the not so distant past when religion reigned supreme, and science and technology played a negligible role in everyday life. Now that the rise of science and technology has enabled enormous increases in the human population, we must face the question: How many people can live on the surface of the Earth?

The rate of growth, that is, the percentage increase per year, of the human population is at an all-time high. Many politicians, economists and religious leaders regard rapid population growth as "natural" and extol its virtues. For example, we often read that population growth stimulates the economy and is, therefore, good. This may be true if one's vision is expressed in units of four years and limited to a few decades at most. However, life has existed on Earth for at least 3.5 billion years and human beings for a few million years. So the very rapid population growth of the past 100 or so years is not natural.

If the current rate of growth of world population, about 2% per year, were continued into the year 3400, then each person now alive would have 1 trillion descendants and the total human population would be about 10,000,000,000,000,000,000,000 (10 sextillion). This is 10% of the total number of stars in the entire observable universe. Well before 3400, the average amount of land per person would have diminished to less than one square inch.

Broken down into shorter intervals, we are talking about a tenfold increase in just over a century. So by about the year 2100, at current rates, there would be 50 billion people on Earth. And 500 billion not long after the year 2200.

What about covering all the deserts and oceans with people? That would only delay the inevitable need for zero population growth by a century at the very most. How about shipping off the extra people (net difference between the number born and the number who die) to outer space? At current growth rates that would mean sending 10,000 people up every hour of every day. And we have trouble launching a few shuttles per year safely.

The issue is not whether an economy that is stimulated by population growth is good, bad or indifferent. The above calculations show that growth is impossible except in the very short run. The real issue is what kind of world will the people of the present and next few generations leave for the people and other creatures of the next few millennia.

According to UCLA professor and biologist Jared Diamond, the coming century will witness one of the worst extinctions in the history of life on Earth. Roughly one-half of the 30 million species that are estimated to exist will become extinct, courtesy entirely of human beings. If yet additional population and economic growth of the sort that some people espouse actually occurs, then the extinction rate will be even worse. A combination of far too many people, greed and unbridled technological power is destroying the natural world.

Each person plays a role in the population equation. If you and your spouse have two children and four grandchildren then you are reproducing at replacement levels (zero population growth). But if you have four children and they, in turn, each have four children so that you have 16 grandchildren, then that is roughly equivalent to the 2% per year growth rate that characterizes the world as a whole.

Each of us has his or her own system of values. For me, a planet with relatively few people, each of whom can live with dignity and a high quality of life, is far superior to a world where too many people, awash in pollution, stretch resources to the breaking point and where billions struggle to survive at mere subsistence levels.

—BEN ZUCKERMAN

BIOGRAPHIES

CHAPTER 1: POPULATION: CHALLENGE TO BIOSPHERE AND BEHAVIOR

Jean-Michel Cousteau

Jean-Michel Cousteau, the son of environmentalist and ocean pioneer Jacques-Yves Cousteau, has spent his life exploring the world's oceans aboard the research vessels *Calypso* and *Alcyone* and communicating to people of all nations and generations his love and concern for the planet. He is an impassioned, eloquent environmental spokesman, lecturing to as many as 100,000 students a year. In the mid-1960s he worked on the award-winning television film series *The Undersea World of Jacques Cousteau* and later served as executive producer of other notable films including *Jacques Cousteau: The First 75 Years*, *Cousteau/ Amazon*, and the Emmy award–winning *Cousteau/Mississippi*. He is the producer of the current television series *Cousteau's Rediscovery of the World*, composed of 35 hour-long documentaries. For over a decade he was executive vice-president of the Cousteau Society, and vice-president of Equipe Cousteau, the society's sister company in France. In 1992, he established Jean-Michel Cousteau Productions to expand the range and depth of educational film programming. He was a member of the selection committee for the International NASA/AIA Space Station design competition in 1972, and is also a member of the advisory board of *Outside* magazine, John Denver's Windstar Foundation, and the International Advisory Board of the Professional Association of Diving Instructors (PADI). He holds an honorary doctorate in Humane Letters from Pepperdine University.

CHAPTER 2: THE GLOBAL WARMING DEBATE: ARE THERE PUBLIC POLICY IMPLICATIONS?

Stephen H. Schneider

Stephen H. Schneider is a professor in the Department of Biological Sciences and the Institute for International Studies at Stanford University, and is a senior scientist at the National Center for Atmospheric Research. He is author or coauthor of the books *Global Warming: Are We Entering the Greenhouse Century?*, *The Genesis Strategy: Climate and Global Survival*, and *The Coevolution of Climate and Life*, as well as of more than 175 other scientific publications. He is also editor of the scientific journal *Climatic Change*. A frequent witness at congressional hearings, he has been a member of the Defense Science Board Task Force on Atmospheric Obscuration, and served as consultant to both the Carter and Nixon administrations. Dr. Schneider is much involved in public education on environmental issues, having appeared on *NOVA*, *Planet Earth*, *20/20*, *The Today Show*, and many other national and international science and news programs. He was selected by *Science Digest* in 1984 as one of the "One Hundred Outstanding Young Scientists in America" and is a Fellow of the American Association for the Advancement of Science. He is recipient of the Louis J. Battan Author's Award from the American Meteorological Society, the AAAS-Westinghouse Award for Public Understanding of Science and Technology, and a MacArthur Foundation Prize Fellowship.

CHAPTER 3:   ON THE SUSTAINABILITY OF RESOURCE USE: POPULATION AS A DYNAMIC FACTOR

John Harte

John Harte holds a joint professorship in the Energy and Resources Group and the Department of Soil Science at the University of California, Berkeley. He is also a senior investigator at the Rocky Mountain Biological Laboratory and faculty senior scientist at the Lawrence Berkeley Laboratory. Dr. Harte has served as a member of the National Council of the Federation of American Scientists, an appointee to the Scientific Advisory Committee of the California Air Resources Board, as president of the board of trustees of the Rocky Mountain Biological Laboratory, and as a member of the board of directors both of the Point Reyes Bird Observatory and of the Denali Foundation. He has served on numerous National Academy of Sciences committees on environmental and educational issues and is an associate editor of *Annual Reviews of Energy and the Environment.* An elected fellow of the American Physical Society, he is recipient both of a Pew Scholars Prize in Environment and Biological Conservation and of a Guggenheim Fellowship. He is the author of five books and more than 100 scientific publications on topics including theoretical elementary particle physics, ecology, energy and water resources, consequences of nuclear war, human and ecosystem toxicology, causes and consequences of climate change, acid precipitation, and biological conservation.

CHAPTER 4:   BIODIVERSITY: WHERE HAVE ALL THE SPECIES GONE?

Mildred E. Mathias

Mildred Mathias, late professor of Botany at the University of California, Los Angeles, had a distinguished career in the study of vascular plants, especially tropical ornamental and medicinal plants, and in promoting conservation of natural areas and genetic resources. She was the author of more than 200 scientific and popular publications and was recipient of many awards and honors, including the Charles Lawrence Hutchinson Medal of the Chicago Horticultural Society, the Los Angeles Times Woman of Achievement Award, and the UCLA 1990 Emeritus Professor of the Year Award. She was an elected fellow of the California Academy of Sciences and the American Association for the Advancement of Science, and was exceptionally active in numerous scientific and conservation organizations, including service as president of the Southern California Chapter of the Nature Conservancy, the Southern California Horticultural Institute, the Botanical Society of America, the Western Society of Naturalists, and the American Society of Plant Taxonomists. She traveled and conducted research throughout the world; on seven occasions, newly discovered plant taxa have been officially named in her honor.

CHAPTER 5:   GENDER BIAS AND THE SEARCH FOR A SUSTAINABLE FUTURE

Jodi L. Jacobson

Jodi L. Jacobson is director of the recently established Washington-based Global Development Policy Project. The project's mission is to explore the gender, health, and human rights dimensions of population and development policies, and to transform U.S. foreign policy and related financial aid toward promotion of truly sustainable worldwide development. The project's first initiative is a Working Group on Reproductive Health Policy made up of representatives of the Population Council, the International Women's Health Coalition, and other major organizations. Before establishing the project, Ms. Jacobson was a senior researcher at the Worldwatch Institute where she authored seven highly influential "Worldwatch Papers," including most recently *Gender Bias: Roadblock to Sustainable Development*. She is coauthor of several books and of eight editions of *State of the World*, the annual publication of the Worldwatch Institute, and has written numerous articles for *World Watch* magazine and op-ed pieces that have been reprinted throughout the world. Ms. Jacobson is currently writing a book tentatively titled *Human Rights, Human Needs, and Human Numbers: Forgotten Dimensions of the Population Debate*.

CHAPTER 6: POLITICS AND SOCIETY: POLITICAL CHALLENGES OF CONFRONTING POPULATION GROWTH

Anthony C. Beilenson

U.S. Congressman Anthony C. Beilenson (Democrat, 24th District, California) has long been a protector of natural resources and the environment. Among his numerous accomplishments, he authored the 1978 legislation creating the Santa Monica Mountains National Recreation Area in southern California. He is also author of the African Elephant Conservation Act of 1988, which catalyzed a major international campaign to save the elephant that resulted in the 1989 worldwide ban on the trade in elephant ivory. Congressman Beilenson has been a leading supporter of international family planning programs and is cochair of the Congressional Coalition on Population and Development. He is a senior member of the Rules Committee, which controls consideration of all significant legislation in the House of Representatives, is a ranking Democrat on the House Budget Committee, and he has served as chairman of the House Permanent Select Committee on Intelligence. Regarded as one of the "20 Smartest Members" of the U.S. Congress in a recent survey, in 1989 he was named by *U.S. News & World Report* as one of the House of Representatives' "Straightest Arrows," an accolade bestowed upon 12 representatives "whose integrity is beyond question."

CHAPTER 7: GLOBAL ENVIRONMENTAL ENGINEERING: PROSPECTS AND PITFALLS

Richard P. Turco

Richard P. Turco is professor of Atmospheric Sciences at the University of California, Los Angeles, where he studies planetary atmospheres and Earth's global climate change. He has published more than 160 scientific papers, articles, and reports, making fundamental contributions to understanding Earth's ozone layer and the "ozone hole" over Antarctica; the climatic effects of volcanic eruptions; the theory of "nuclear winter"; mass extinctions related to meteor impacts; and the global effects of aircraft and rocket engine emissions. Professor Turco is president of the Atmospheric Sciences section of the American Geophysical Union, is recipient of a MacArthur Foundation Prize Fellowship and of the Leo Szilard Prize for Physics in the Public Interest, and has twice been honored by his selection as the H. Julian Allen Awardee of the American Physical Society. With Carl Sagan, he coauthored the influential and thought-provoking volume *A Path Where No Man Thought: Nuclear Winter and the End of the Arms Race*. He is currently completing a textbook to be published by Oxford University Press entitled *Earth under Siege: Air Pollution and Global Change*, and he is actively involved in organizing the geoscience community to confront the growing dangers posed by "global engineering" schemes that seek to cure such planetary environmental problems as greenhouse warming.

CHAPTER 1

POPULATION: CHALLENGE TO BIOSPHERE AND BEHAVIOR

■

Jean-Michel Cousteau*

■

INTRODUCTION

My window on the environment opened when I was seven years old. It was a smallish window about the size of a dive mask. In fact, it *was* a dive mask. Through it, I explored a world that few people had ever seen before, a world of incredible creatures beyond the imagination of a small boy, a tranquil realm of liquid and light.

It was also through that window that, over the following decades, I watched what happened to that undersea world that remained out of sight and out of mind to a population that was in the process of doubling within a generation. Pollution and depletion, the declining health of pristine areas, inspired in me the desire to protect what was left of the world I loved.

Once, environmental protection meant maintaining wide open spaces, the great outdoors, scenic vistas, and wildlife. Later, with more sophisticated understanding, environmental protection came to mean preserving the delicate mechanisms of air, ocean, water, and land that make life possible on Earth. Today, however, environmentalism can no longer exclude human population dynamics, and the social and economic conditions that are the wellsprings of overpopulation.

Human population growth is the primary environmental challenge we face, not because it is by itself more serious than radioactive waste disposal, dioxins, or the loss of biodiversity, but because it is hard to envision a solution to these and most other problems without solving the population problem first.

Further, many of the challenges we face, such as resource depletion, air and water pollution, or desertification, are themselves directly or indirectly initiated by population pressures. We must investigate these relationships if we are to modify our behavior in a way more befitting the limitations of life on this planet.

Population and environment form an enormously complex web of issues. Yet, these many complexities are woven together by one simple thread, one common assumption that under-

*Jean-Michel Cousteau Productions, 1933 Cliff Drive, Santa Barbara, CA 93109

1

Figure 1.1

Street scene in Agra, India, a country challenged by a large and rapidly growing population. (Photograph courtesy of R.C. Murphy)

scores why we are in a mood of acute concern: Increasing human numbers means declining quality of life.

It is appropriate that we address this issue within an evolutionary framework. Overpopulation is more than the correlation of human numbers and certain quantifiable levels of environmental degradation; it represents both a crisis in human cultural institutions and a challenge to the human potential for rapid cultural evolution to keep pace with a drastically changed environmental situation.

Once our numbers would have been cause for considerable pride, a tangible expression of our ability to so master the resources of our world that we could breed with impunity and never undermine our own chances for survival. Our reproductive success is prodigious for a species with few natural advantages beyond our brain and our hands. In fact, however, our success is fairly recent, and driven, on the positive side, by artificial inputs such as modern medicines and chemical agriculture, and on the negative side by poverty and the low status of women. Today's population dynamics erode the positive inputs while reinforcing the negative stimuli.

THE PATH OF LEAST PRECAUTION

It took hundreds of thousands of years for the human population to reach 2 billion after the Second World War. Today we stand at 5.6 billion. Based on 1990 world population and the numbers of women who are or will soon be in their childbearing years, the United Nations projects that at today's fertility rates, the world population could reach 6.2 billion by the year 2000 and 8.5 billion by the year 2025. Even these estimates are based on already falling fertility rates, for despite a drop in the global population growth rate from 2 to 1.6 percent per year, the population is still growing at the rate of 90 million people per year (Population Action International, *Challenging the Planet*, 1993).

Like all other animal species, human beings consume resources and produce waste material. Because we live at the top of the predatory hierarchy, we tend to consume more and pollute more than other species. And because we have become so numerous in so short a time span, that cycle of consumption and pollution has a hugely destructive impact on the ecosystems that sustain us. So it is not surprising that the chief environmental consequences of overpopulation are scarcity and contamination.

During the Cousteau expedition to Haiti in 1985, we were able to examine a microcosm of population pressure: 6 million people crowded into that tiny country, rainforests chopped down for firewood, fisheries depleted, and reefs dying. Poverty and political instability join population growth in a feedback loop of despair. For many people, the only alternative is to flee.

The seas, from which life emerged, and a memory of which remains chemically constituted in our very blood, are even in their immensity seriously influenced by human population growth. Increased demand, technological advances, and poor management have combined to bring the world's **fisheries** near the brink of destruction. From 1950 to 1987, the world fish catch rose from 22 million tons to 92 million tons (*Calypso Log*, 1991). According to the Worldwatch Institute (1985), 11 major fisheries had been depleted to the point of collapse, including Atlantic herring, Alaska king crab, and the North Atlantic cod. In order to protect their ravaged fisheries, nations have pushed limits to 200 miles and beyond. But this does not affect passive methods such as monofilament gill-netting. According to a just-released report of the Center for Marine Conservation, a typical gill-netting vessel may set up to 40 miles of net daily. In the North Pacific, gill netters may use over 2 million miles of netting yearly in the search for tuna and squid—enough to circle the Earth 88 times (Norse, 1993)! Almost half of their catch are "nontarget" species, such as dolphins and sea birds. Their carcasses are dumped back into the sea, where, alive, they bolster the diversity of the sea. Dead, they are useless to the oceanic ecosystem and each other. In many regions, such as Mexico's Sea of Cortez, where an expedition brought us in 1985, overfishing has so reduced the population of sharks and shrimp that other, less palatable species such as manta rays are now the catch of last resort.

Land-based activities that use the sea as a dumping ground for industrial and chemical wastes are already supremely threatening to the sea, especially locally. Population growth not only compounds the effect, but does so progressively when we consider the additional impact of overfishing.

The world's **forests** are under pressure from population growth both directly and indirectly: directly through the worldwide shortage of firewood and the clearing of land to make way for resettlement projects, and indirectly through industrial practices that result in acid rain, and the callous misuse of tropical hardwoods (Park, 1992).

It has also been apparent for a decade that many of our cherished industrial practices, on the ever-expanding scale we have envisioned for them, may have damaged our atmosphere in a way we have yet to understand fully. Inexpensive energy is essential to the industrial way of life. So far we have favored **nonrenewable energy sources** such as the burning of coal and oil (the effects of hydroelectric and nuclear power are other problems altogether). The burning of fossil fuels has raised carbon dioxide levels in the atmosphere substantially in the last 40 years (PAI, *Challenging the Planet*, 1993).

We are not only numerous, but we are also crowding other species out of existence, both with our sprawl and through the activities made necessary by our various life-styles. The rate of **species extinction** around the world is accelerating as habitats are destroyed. Currently, more than 27,000 animal and plant species suffer extinction every year. Much of the time, steady encroachment of human numbers, rather than cataclysmic events, causes a species to vanish. Even putting a road through a rainforest may be enough to critically disturb the range of a particular animal, alter the animal's behavior, perhaps even affect its reproductive success. Many plants have coevolved with particular species of animals that they depend on for pollination; once the animal is removed from the web of life, the plant will follow (Park, 1992).

The loss of a single species may change the biosphere in ways we only dimly perceive, much less understand. Ecosystems that lose the services of certain species are less able to maintain and regenerate themselves. By indiscriminately removing one species, we do more than just favor its competitors and its prey—itself a potentially undesirable consequence, as we have evolved within a matrix of presently existing species—we lose all the information encapsulated in its genetic codes. As we have learned from ethnobotanists in recent years, much of what we are destroying is of immense value to us nutritionally and medicinally.

As polluters, we thus set in motion a domino effect that ultimately rebounds to our disadvantage as consumers. The environmental consequences of overpopulation are ultimately the source of human suffering and thus are inseparable from the human situation: an ever more difficult struggle for survival with fewer and fewer tools.

My focus for many years has been water, the life's blood of our planet and our species. In the beginning, our expeditions centered on the seas. However, as we became more aware of human impacts on the water cycle, we began to focus on rivers, the interface of human activities, and, ultimately, the world's oceans. The result of these journeys along the Amazon, Sepik, Mekong, and others, is a sensitivity to the threat that water systems in particular face in the light of overpopulation.

Overpopulation threatens **water quantity and quality.** Usable water is scarce to begin with: Only about 2.5 percent of all the water on Earth is so-called fresh water, and 70 percent of this is locked up in polar ice caps. Human use of fresh water has increased four times since 1940, to the point that existing sources must now be considered fully exploited. In many cases, water is used unsustainably, with insufficient amounts left in rivers for the river to perform necessary functions. Conservation represents what amounts to a new source, but will it be enough? Not if population continues to grow. More than 1.2 billion people currently do not have access to safe, clean drinking water. More than 40 percent of the world's population in 80 countries suffers from chronic water shortages. This percentage is even higher in countries with high birth rates. In Guinea and Angola, for example, between 60 and 70 percent of the population are without access to safe drinking water. Furthermore, the outlook is not positive: Of the 20 countries in which water was considered scarce in 1990, 15 had growing populations (PAI, *Sustaining Water*, 1993).

With more and more people flowing into the mega cities of the Third World, urban infrastructures are stressed beyond their limits. How can Mexico City, for example, hope to keep up with a population that has grown from 1 million to 20 million in only 50 years? Where it is available, water is increasingly polluted. Mostly, this is due to human waste— raw sewage. Once again, this impacts the ability of people to survive. In Jakarta, Indonesia, the waste of 7.5 million people flows directly into Jakarta Bay, contaminating clam and oyster beds that are important for the local economy. In the fast-growing cities of West Africa, freeways and skyscrapers are only a mask of prosperity, while hepatitis, cholera, typhoid, and dysentery stalk the waterways. According to Population Action International, half of the world's population suffers from water-related diseases (PAI, *Sustaining Water*, 1993).

In the countryside, attempts are being made to bring every available acre under cultivation. Rainforests are slashed and burned for a few good seasons of pasture or farming on poor tropical soils. Abundant water is lavished on crops, together with the petroleum-based chemicals that are needed to boost production to fulfill the basic needs of our population. Erosion soon turns this dream to dust. Not only is topsoil lost as runoff during tropical rains, but also river and estuary life is choked. Coral reefs become unproductive, thus sharpening the agony of food and resource shortages. Where chemicals are heavily used, they are washed into the brew as well. This nonpoint source pollution also seeps into the water table, contaminating ground water before it is even pumped.

Adequate treatment of available water is a luxury that few societies can afford. Even in the United States the federal government has quietly withdrawn from the financing of waste water treatment plants, placing a huge burden on local budgets. Meanwhile, nonpoint source contamination such as runoff from city streets and agricultural activities overwhelms the best efforts of coastal municipalities to maintain high water quality.

Because of water scarcity, overpopulation looms as a major cause of unrest and international friction. During a Cousteau expedition to Haiti we watched as hundreds of people gathered in the predawn hours with pails and pitchers to fight each other for access to a public spigot. Desperately kicking and shoving, only a few of these people got what they came for, while others, too weak or timid, were obliged to scoop water from the gutter, or from the canals that serve as communal washbasins and sewers.

On the international level, water has played a large role in the recent wars in the Middle East. High birth rates have forced such countries as Turkey, Jordan, Egypt, and Iraq to consider and execute vast reclamation schemes, which divert water from neighboring states, causing tension and violence. Water was one of the root causes of the Six-Day War and remains one of the chief points of discussion between Israel and Jordan today. Huge reclamation projects also increase the overall debt of developing countries, which means even more environmental corners are cut in other sectors in a futile effort to keep inflation under control.

Population growth drives the desperate search for quick fixes to economic hardship. In Africa, more and more marginal land is being cultivated, land that has no business being farmed. It quickly erodes and is useless even for grazing. According to the Global Assessment of Soil Degradation, 17 percent of the land in use is degraded (Worldwatch Institute, 1994). Not only does this threaten world food production, but it has also been linked to changes in local climate, altering the ability of entire ecosystems to regenerate themselves.

Even our fixes have **problematic side effects.** Massive quantities of nitrates, phosphates, and petroleum-based chemicals have been used since the Second World War to spur food production, resulting in the agricultural boom known colloquially as the "green revolution." Now, however, the soils are becoming exhausted and infertile. Pesticides can no longer keep up with the mutation rate of harmful insect species, resulting in larger and more toxic doses that find their way into the human body. And salinity, the eternal enemy of hydraulic cultures from Mesopotamia to modern Kansas, has also become acute around the world wherever irrigated agriculture is practiced beyond the carrying capacity of the land.

■

CONSUMING THE FUTURE

For many observers, overpopulation is an issue that applies only to the developing world. Perhaps because the environment we have inherited was at that time already seriously degraded, we are shocked by the images that flow back to us from the so-called **Third World:** famine, disease, and rainforest destruction.

Yet, there is a population problem in the **industrialized world** as well, and it is itself one of the factors that worsens the situation for the rest of humanity. Overconsumption of resources is the population dilemma of the richer nations. According to Worldwatch Institute researcher Alan Durning, the inhabitant of an industrial society consumes 3 times as much fresh water, 10 times as much energy, and 19 times as much aluminum as someone in a developing country. He or she also generates 96 percent of the world's radioactive waste and 90 percent of the chlorofluorocarbons that eat away at the ozone layer (Durning, 1992).

Despite their harmful impact, the economies of the industrialized nations set the tone and shape the aspirations of the other 4.5 billion people on Earth. Televisions bring the gospel of consumerism to the farthest reaches of the planet. Yet ironically, our prosperity rests in large degree on an economic system that so far has deprived the rest of the species of fulfilling its dreams. Debt, resource exploitation, and capital transfer lock the developing world into vicious cycles of poverty. For many people in those societies, reproduction is the most obvious way to increase the domestic labor force and ensure that one is taken care of in one's old age.

Social inequities become more pronounced once the downward spiral begins, and it is often difficult to assess which came first, deprivation or population growth. The inferior status of women, however, is one of the most easily identifiable factors linking social institutions, poverty, population growth, and environmental decline. Wherever women do not have access to education or control over the number of children they bear, high birth rates are prevalent. This trend only intensifies the initial impoverishment. Pressure on the family immediately becomes pressure on the urban infrastructure, the overworked land, the forest, the sea.

Thus far, the introduction into the developing world of high-tech tools of advertising, high finance, and modern communication has done little to improve the status of women or stem the tide of population growth. Rather, the developing world absorbs new technologies while maintaining cultural traditions and economic tendencies that lock population growth in place.

The Earth already cannot support even the one billion or so Western consumers it now hosts, to say nothing of an expansion of this life-style among the rest of the species. Our prosperity has many hidden environmental costs, which until now we have been able to cover with ever more ingenious and massive inputs of energy and capital. The bill is coming due in the industrialized world, where we are beginning to understand the connections between pollution and diminished prosperity, but for the very poor on this planet, those still so far from being able to meet even their most basic needs, the hidden costs of industrialization seem trivial compared to the imperatives of survival.

THE EVOLUTIONARY CHALLENGE

Ultimately, the population's environmental impact is an issue of **human ethics.** Population growth is the symbol of a species no longer cognizant of the natural systems that sustain it. Current human assumptions and institutions undermine the physical foundations on which all societies rest. Our consensus that population is the root of environmental degradation is certainly compelling, but it is not the whole story. Desertification and rainforest depletion, for example, are indeed population driven, but they are also a result of inequities in land distribution. Only by looking more deeply into the institutional crisis—the cult of the individual, the myth of eternal economic growth, the widespread ignorance of basic biological principles, the breakdown of cohesive human groups by industrial processes—can we begin to develop values more in touch with the demands of our physical situation. Thus, just as environmental ills oblige us to seek their origins in human numbers, population growth forces us to look at environmental issues in the widest possible intellectual context.

The issue is not preserving the environment for its own sake. The planet will continue to orbit the sun, no matter what we humans may choose to do about our population, whether we poison and deplete the biosphere or not. The only relevant task for us is to define what we consider essential to our own quality of life.

My travels have impressed on me how different cultures traditionally define quality of life in terms of their own priorities and historical experience. However, a common thread seems to be the imperative of leaving future generations with a certain range of survival options. Integral to these options is a functioning ecosystem that can sustain those that must continue to depend on it for survival. It might be a certain patch of rainforest or a flock of sheep or a healthy coral reef, coupled with the knowledge to steward it wisely.

If quality of life is defined as the ability to develop within the widest possible range of options, population growth is fueling a precipitous decline in quality of life for most people on the planet. We have now attained a situation whereby the numbers of people affected by poverty and disease may even recede here and there as a percentage of the overall total, but the numbers of people added every day make even our successes seem meaningless.

We have seen how population pressures damage the health of the biosphere and undermine quality of life for vast numbers of people. We must also note, however, that the problem is also one of human attitudes and convictions and, as such, it is reversible. Yet any optimism we may feel should be infused with urgency. The likelihood that such a reversal in our behavior will have any chance of bettering the human condition diminishes the longer we hesitate to address the cultural, economic, and social origins of population growth.

■

REFERENCES

Calypso Log, February, 1991. (Chesapeake, Va.: Cousteau Society).

Durning, A. 1992. *How Much Is Enough?* (New York: W.W. Norton).

Norse, E.A. (ed.). 1993. *Global Marine Biological Diversity* (Washington, D.C.: Center for Marine Conservation/Island Press).

Park, C. 1992. *Tropical Rain Forests* (London: Routledge).

PAI (Population Action International). 1993. *Challenging the Planet* (Washington, D.C.: Population Action International).

_____. 1993. *Sustaining Water* (Washington, D.C.: Population Action International).

Worldwatch Institute. 1985. *State of the World,* Brown, L.R., et al. (eds.) (New York: W.W. Norton).

_____. 1994. *State of the World,* Brown, L.R., et al. (eds.) (New York: W.W. Norton).

CHAPTER 2

THE GLOBAL WARMING DEBATE: ARE THERE PUBLIC POLICY IMPLICATIONS?

■

Stephen H. Schneider*

■

THE MEDIA DEBATE

"Global warming" hit the headlines in North America in the wake of the heat wave and fires of 1988. By now, so much misinformation about the **"greenhouse effect"** has been circulated that public understanding is often confused and public policy making paralyzed. The airwaves and printed pages have been clogged with assertions and counterassertions of opposing advocates with charges and countercharges over the alleged seriousness or triviality of global warming. This chapter provides perspective on whether the possibility that human activities may alter the climate has significant public policy implications. First, however, it is necessary to review the debate, including its physical, biological, and social/scientific interactions.

As a climatologist identified with this subject, I am constantly asked to explain what is actually happening and how important it is. Debate has accelerated since 1988 along with a surge of criticism, some well intended, some pure vitriol. The experience since those hotter-than-usual summers of 1987 and 1988 has confirmed for me two crucial points: (1) the extent of public concern ought to be shaped by the best scientific and economic knowledge available about possible long-term climate change and its effects on, for example, farms, floods, sea levels, forest fires, ecosystems, and cardiovascular or tropical diseases; and (2) public concern is being shaped by the blurring of scientific and economic knowledge under the impact of political opinion, media miscommunication, and a polarized debate among battling scientists.

*Department of Biological Sciences and Institute for International Studies, Stanford University, Stanford, CA 94305-5020, and the National Center for Atmospheric Research, Boulder, Colorado 80803. Studies carried out at the National Center for Atmospheric Research are supported by the National Science Foundation; any opinions, findings, conclusions, or recommendations expressed in this article are those of the author and do not necessarily reflect the views of the National Science Foundation. This chapter is adapted in part from S.H. Schneider (1995). The future of climate: Potential for interaction and surprises, in T.E. Downing et al. (eds.) *Climate Change and World Food Security* (New York: Springer) and in part from S.H. Schneider (1993) Degrees of certainty. *National Geographic Research & Exploration* 9: 191–200.

Scientists too often share responsibility with the media for not communicating complex scientific issues clearly to the public. Most members of the general public, as well as many officials in government, do not recognize that most scientists spend the bulk of their time arguing about what they do not know. We simply have to spend more time making clear the distinctions among (1) what is well known and accepted by most knowledgeable scientists, (2) what is known with some degree of reliability, and (3) what is highly speculative.

The public debate on global warming rarely separates these components, thereby leaving the false impression that somehow the scientific community is in overall intellectual disarray on every aspect. In fact, the 15-year-old, often-reaffirmed U.S. National Academy of Sciences (NAS) consensus estimate of 1.5°C to 4.5°C global average warming if carbon dioxide (CO_2) were to double its atmospheric concentration still reflects the best estimate from a wide range of current climate models (IPCC, 1992) and ancient climatic eras (Lorius et al., 1990; Hoffert and Covey, 1992). The Earth has not been more than 1°C to 2°C warmer than now during the 10,000-year era of the development of human civilization. The previous ice age, in which mile-high ice sheets stretched from New York to Chicago to the Arctic and across northern Europe, was only 5°C colder on a global average than the current 10,000-year-old interglacial epoch we now enjoy. This 1.5°C to 4.5°C warming consensus includes recent studies that halved the "best guess" on warming from over 4°C to 2.5°C. Perhaps some new discovery will push it back up again, but in any event, most knowledgeable experts accept the 1.5°C to 4.5°C estimate of warming.

Changes of this magnitude at global scale could dramatically alter accustomed climatic patterns, affecting agriculture, water supplies, disease patterns, ecosystems, endangered species, severe storms, sea level, and coastal flooding. However, all aspects must be stated in terms of subjective probabilities, as very few scientists, myself included, would say they believe the future climate will be in or out of the 1.5°C to 4.5°C warming range for certain. Rather, most scientists believe this range to be reasonably probable (for example, see Table 2.1). Therefore, if scientific opinion is to be communicated accurately, it must be by conveying issues in **probabilistic terms**—even if those probabilities are subjectively rather than objectively determined. The reason that subjectivity is unavoidable is that all assessments of the likelihood of particular climate change scenarios depend on many assumptions, each of whose validity is debated. The media should focus on providing perspective on the subjective probability estimates of a broad cross-section of knowledgeable scientists rather than on conducting an entertaining but misleading debate among the most extreme of the dueling scientists—occasionally stretched beyond caricature in editorials or articles by polemicists and ideologues. What counts, then, is the nature of the evidence and the spectrum of opinions of a broadly representative group of experts (for example, reports of the Intergovernmental Panel on Climate Change [IPCC] or the NAS), not a few highly polarized, visible debaters who receive most of the media attention.

WHAT DOES COMPRISE A CONSENSUS ON GLOBAL WARMING?

To support the contention that much is known and accepted by the vast majority of the knowledgeable scientific community, I offer the following list of a dozen points that are related to global warning. A good source of relevant discussion is the National Research Council's study on global warming and its implications (NAS, 1991). (The bracketed comment after each statement is my estimate of the likelihood of the statement's being valid.)

1. **Greenhouse gases** such as water vapor (H_2O), carbon dioxide (CO_2), methane (CH_4), nitrous oxide (N_2O), and chlorofluorocarbons (CFCs) trap infrared radiative energy in the lower atmosphere. [Certain]

Table 2.1

Executive summary of chapter 5 of the Intergovernmental Panel on Climate Change (IPCC, 1990).

1. All models show substantial changes in climate when CO_2 concentrations are doubled, even though the changes vary from model to model on a sub-continental scale.

2. The main equilibrium changes in climate due to doubling CO_2 deduced from models are given below. The number of *'s indicates the degree of confidence determined subjectively from the amount of agreement between models, our understanding of the model results and our confidence in the representation of the relevant process in the model. Five *'s indicate virtual certainties, one * indicates low confidence.

Temperature:
*****   the lower atmosphere and Earth's surface warm;
*****   the stratosphere cools;
***   near the Earth's surface, the global average warming lies between +1.5°C and +4.5°C, with a "best guess" of 2.5°C;
***   the surface warming at high latitudes is greater than the global average in winter but smaller than in summer. (In time dependent simulations with a deep ocean, there is little warming over the high latitude southern ocean);
***   the surface warming and its seasonal variation are least in the tropics.

Precipitation:
****   the global average increases (as does that of evaporation), the larger the warming, the larger the increase;
***   increases at high latitudes throughout the year;
***   increases globally by 3 to 15% (as does evaporation);
**   increases at mid-latitudes in winter;
**   the zonal mean value increases in the tropics although there are areas of decrease. Shifts in the main tropical rain bands differ from model to model, so there is little consistency between models in simulated regional changes;
**   changes little in subtropical arid areas.

Soil moisture:
***   increases in high latitudes in winter;
**   decreases over northern mid-latitude continents in summer.

Snow and sea-ice:
****   the area of sea-ice and seasonal snow-cover diminish.

The results from models become less reliable at smaller scales, so predictions for smaller than continental regions should be treated with great caution. The continents warm more than the ocean. Temperature increases in southern Europe and central North America are greater than the global mean and are accompanied by reduced precipitation and soil moisture in summer. The Asian summer monsoon intensifies.

3. Changes in the day-to-day variability of weather are uncertain. However, episodes of high temperature will become more frequent in the future simply due to an increase in the mean temperature. There is some evidence of a general increase in convective precipitation.

4. The direct effect of deforestation on global mean climate is small. The indirect effects (through changes in the CO_2 sink) may be more important. However, tropical deforestation may lead to substantial local effects, including a reduction of about 20% in precipitation.

5. Improved predictions of global climate change require better treatment of processes affecting the distribution and properties of cloud, ocean-atmosphere interaction, convection, sea-ice and transfer of heat and moisture from the land surface. Increased model resolution will allow more realistic predictions of global-scale changes, and some improvement in the prediction of regional climate change.

2. The natural greenhouse effect from clouds, water vapor, CO_2, and methane is responsible for some 33°C (60°F) of natural surface temperature warming. [Virtually certain]

3. Humans have altered the natural greenhouse effect by adding 25 percent more CO_2, 100 percent more CH_4, and a host of other greenhouse gases such as N_2O and CFCs since the Industrial Revolution. [Virtually certain]

4. Added greenhouse gases from human activities since preindustrial times should have trapped some two watts of infrared radiative energy over every square meter of Earth. This is well established based on our considerable knowledge of the structure of the atmosphere and extensive validation of relevant processes from satellites and other measurements—even though the extra two watts cannot be directly measured yet. [Very likely]

5. The Earth has, in fits and starts, warmed by about 0.5°C over the past century. The 1980s are the warmest decade on record, and 1990, 1991, and 1988 were (in order) the warmest years in the thermometer records. [Very likely]

6. At the highly significant, often-cited 99 percent statistical confidence limit, correlations between the observed warming and the buildup of human-induced greenhouse gases cannot be asserted for at least another decade or two. Nonetheless, the likelihood that the 0.5°C 20th-century warming trend is wholly a natural phenomenon is small (I would estimate perhaps a 10 to 20 percent chance—Schneider, 1994). [Likely]

7. Most climatic models project a warming of several degrees or so in the next 50 years given standard greenhouse gas emission scenarios, and they portend a potential long-term (that is, 2100 A.D. to 2200 A.D.) warming commitment as high as 5°C to 10°C (for example, IPCC, 1992). [Good chance, perhaps an even bet]

8. Natural, sustained, globally averaged rates of surface air temperature change (for example, from the breakup of the last ice age 15,000 years ago to the full establishment of our current interglacial age some 5,000 to 8,000 years ago) are about 1°C per 1,000 years. On the other hand, even the minimum projected human-induced rates of global climate change are on the order of 1°C per 100 years up to a potentially catastrophic rate of 5°C per 100 years—the latter being some 100 times faster than typical sustained globally averaged rates of climate change during which human civilization evolved and from which the current distribution of species and ecosystems emerged. [Very likely]

9. Different species (for example, specific kinds of trees, insects, birds, and mammals) would all respond differently to projected climatic changes. For example, birds can migrate rapidly, but vegetation some birds need for survival habitat would respond only very slowly (over centuries). This implies a tearing apart of the structures of communities of plants, insects, and animals (as discussed, for example, in Root, 1992) at rates that exceed most clear prehistoric or geologic metaphors (Graham and Grimm, 1990). [Very likely]

10. Most forest species migrate at rates of some one kilometer per year, and many species would not be able to keep pace with temperature changes at the rate of several degrees centigrade per century without human intervention to transplant them (that is, ecological engineering). [Likely]

11. Current engineering and economic practices in terms of building standards, automobile mileage, power production, or manufacturing are very retarded relative to the energy efficiency of best available technologies or techniques. Many studies (for example, NAS, 1991, and OTA, 1991) suggest that from 10 to 40 percent reductions in current CO_2 emissions in the United States could result in averaged costs at or below current rates of expenditure for the equivalent energy services if current inefficient practices/infrastructures were replaced by state-of-the-art, proven efficient practices/equipment. [Likely]

12. Despite the slowness (approximately 1°C per 1,000 years) of sustained globally averaged natural roles of climatic changes over most of the past 160,000 years, there are a few examples of very rapid changes seen in pollen records in Europe (Woillard and

Mook, 1982) and ice cores in Greenland (GRIP, 1993). The possibilities of such "surprise" scenarios arising from greenhouse gas emissions, while currently hard to assess, are sufficiently dramatic as to require both scientific and policy assessments. [Speculative, but urgent]

The uncertainties in temperature projections over the next century range over a factor of 10 and are well summarized by Figure 2.1. This is an attempt to include uncertainty from human behavioral activities that create greenhouse gas emissions, biological factors that influence the carbon cycle and thus CO_2 and CH_4 concentrations, and physical factors such as the **"feedback effects"** of clouds or ice, all of which taken together lead to the wide differences seen in Figure 2.1 (Jager, 1988).

■

WHAT IS KNOWN WITH SOME RELIABILITY

A major criticism of global warming concerns the apparent discrepancy between the erratic warming of the Earth and the relatively smooth increase in greenhouse gases over the past hundred years (Figure 2.2). It has been alleged that temperature trends in the 20th century cannot be attributed to greenhouse gas buildup, because most of the warming took place between 1915 and the 1940s, followed by a Northern Hemisphere cooling at the very time the global greenhouse gases began to build up rapidly. Then, from the mid-1970s to 1992 there was a dramatic warming, with the last 12 of those years containing over a half dozen of the warmest years on record.

This problem of cause and effect is akin to a criminal investigation in which the whereabouts of one principal suspect is fairly well known, but the whereabouts of other possible secondary suspects were not carefully observed. In this case, of course, the "crime" is the 0.5°C warming trend of the 20th century, and the known principal "suspect" is the increase in greenhouse gases. Unfortunately, we cannot rule out some possible role for the unwatched "suspects," since we do not have accurate ways of measuring precisely what these suspects did. These other potential climatic influences, or **"forcings"** as they are called, include

Figure 2.1

Three scenarios for global temperature change from the present to the year 2100 derived from combining uncertainties in future trace greenhouse gas projections with those of modeling the climatic response to those projections. Sustained global temperature changes of more than 2°C (3.6°F) would be unprecedented during the era of human civilization. The middle to upper range of these scenarios represents climatic change at a 10 to 100 times faster pace than sustained natural global average rates of change (Jager, 1988).

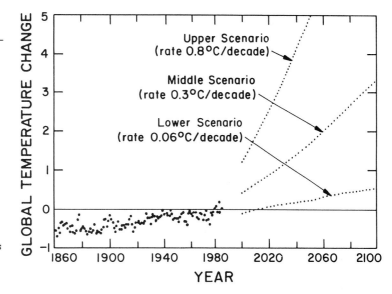

Figure 2.2

Observed global mean temperature changes from 1861 to 1989 compared with predicted values. The data plotted in the solid line showing observed changes, presented in Section 7 of the report prepared in 1990 by the Intergovernmental Panel on Climate Change (IPCC), have been smoothed here to show the decadal and longer time-scale trends more clearly. Predicted changes, shown by the dashed lines, are based on observed concentration changes and concentration/ forcing relationships as discussed in Section 2 of the 1990 IPCC report, calculated by use of an upwelling-diffusion climate model (Wigley and Raper, 1991).

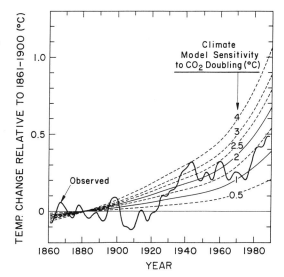

sunspot activity and atmospheric particles from volcanic eruptions, industry, automobiles, and agriculture. It has long been known that most of these particles, for example, tend to cool the planet, counteracting some of the greenhouse effect, at least regionally.

Recently, Charlson et al. (1991) picked up on this old debate (see, for example, Charlson and Pilat, 1969; Schneider, 1971; SMIC, 1971) of the cooling potential of SO_2 emissions (largely from burning sulfur-contaminated oil or coal) and added some quantitative insights. They concluded that sulfuric acid or other **sulfate aerosol particles** (a form of smog) could both directly and indirectly (by brightening clouds) reflect enough sunlight away so as to nearly compensate for the extra ground-level heating, from human-caused greenhouse gases (CO_2, CH_4, N_2O, and CFCs) over most Northern Hemisphere land masses since the 1960s. However, because reflected sunlight is a daytime phenomenon whereas the addition of greenhouse gases is a day-and-night effect, some scientists (for example, Kerr, 1992) recently have begun to suspect that the SO_2 effect combined with the anticipated global warming from greenhouse gas emissions would, at least over land in the Northern Hemisphere, result in a nighttime warming trend. Karl et al. (1992) noted that over the United States, the former Soviet Union, and China (precisely those places most affected by SO_2 emissions), warming trends during the past 30 years were indeed largely at night. Although 30 years is too short to enable statistically secure conclusions, and radiative effects on the day/night cycle of regionally patchy aerosols require sophisticated calculations yet to be made, nevertheless, these results add (not subtract, as some critics contend) to the confidence that greenhouse gas buildups equivalent to a doubling of CO_2 will eventually warm the Earth by 1.5°C to 4.5°C. This is noted in the interim report of the Intergovernmental Panel on Climate Change (IPCC, 1992) as the likely warming during the next 50 or so years, although most scientists agree that establishing this securely requires 10 to 20 more years of temperature, solar, atmospheric pollution, and volcanic measurements.

One final aspect needs mention. We should take little comfort from the possibility that sulfuric acid particles will "save us" from global warming for two reasons. First, such chemicals are a principal ingredient of acid rain and health-threatening smog. Second, aerosols are, as many have noted for decades (for example, Schneider and Mesirow, 1976), a regional phenomenon, whereas greenhouse heat-trapping effects are spread fairly uniformly over the globe. Thus, even if on a hemispheric average sulfur aerosols were to reject exactly as much extra solar heat to space as greenhouse gases trap heat at infrared wavelengths near the surface, this would not cancel competing climatic effects, because cooling would be in patterned, half-continental-sized patches whereas heating would be more evenly distributed around the Northern Hemisphere. The likely result would be distortion of normal heating patterns, such as the land/ocean thermal contrast. Such distortions would likely lead to regional climatic anomalies (that is, unanticipated local/regional climatic events) even if the net hemispheric temperature changes were small as a result of the hemispheric-scale heating/cooling compensations (Schneider, 1994). In short, we cannot "cure" global warming with sulfur dioxide emissions and escape risk-free.

Actually, evaluating the significance of global climate signals attributed to human activities (Solow and Broadus, 1989) is not objective: The assignment of formal statistical significance requires two fundamentally subjective judgments about: (1) the long-term (that is, low frequency) natural climatic variability, and (2) the time patterns of both natural and anthropogenic forcings (such as greenhouse gas buildups or changes in the solar output or aerosols).Thus, rather than delude ourselves by relying on some arbitrary statistical confidence limit, we should recognize that most probabilities assigned to specific environmental consequences are based merely on what different analysts regard as plausible underlying assumptions. The most reliable intuition integrates the expertise of people from wide-ranging fields who are familiar with many aspects of the problems and their interconnections (for example, NAS, 1991).

In this context of uncertainty but not implausibility, a prudent approach is to analyze events with relatively low or unassignable probability. This is done by nearly all defense establishments and in medical practice where relatively low or unknown probability scenarios with significant potential impact are analyzed. Sometimes major public or private investments are made to hedge against or accelerate the onset of such scenarios (Morgan, 1993).

What is the basis for the moderate degree of certainty (40 to 90 percent) typically assigned to the likelihood of significant climatic change? I believe this substantial consensus arises largely from established knowledge of heat-trapping processes and from validation exercises that test models against features of the present and past climate. In fact, many aspects of these models have already been validated to a considerable degree, although not to the full satisfaction of any responsible scientist.

For example, we know that the last ice age, which was about $5°C$ ($9°F$) colder on a global average than the present era, had CO_2 levels about 25 percent less than the current **interglacial period** (before the Industrial Revolution). Methane, another very potent greenhouse gas, also was lower by about half in the ice age relative to interglacial preindustrial levels.

Ice in Antarctica contains gas bubbles that record atmospheric composition going back more than 160,000 years. Cores drilled into the ice sheets show us that the previous interglacial warm age, some 120,000 to 140,000 years ago, had CO_2 and methane levels comparable to, and temperatures comparable to, or perhaps up to $2°C$ warmer, than those in the present interglacial period.

The well-correlated change in these greenhouse gases and in planetary temperature over geological epochs provides an empirical way to estimate the sensitivity of climate to greenhouse gas concentration changes. Such studies find geological-scale temperature changes from greenhouse gas variations roughly of the magnitude that one would expect based on projections from today's generation of computer models (Hoffert and Covey, 1992; Lorius et al., 1990). However, we still cannot assert that this greenhouse-gas/geological-temperature coincidence is proof that our models are quantitatively correct, since other factors were operating during the ice age/interglacial cycles. Furthermore, leads and lags into and out of ice ages are not consistent. Therefore, the best we can say is that the evidence is strong but circumstantial.

One point related to the ice age/interglacial cycles may be useful here. Each warm interglacial epoch typically lasts 10,000 years. Since our current interglacial epoch is now about 10,000 years old, some experts have suggested that global warming could be "good" as it will hold back the next ice age. However, this view fails to consider that the time frame for completion of natural interglacial-to-glacial transitions (including major ice buildups) is tens of thousands of years, whereas the potential for global warming of 1°C to 10°C is only a century or two—a radical rate of change relative to most sustained, natural global climate changes in geological history. Another difficulty is the discovery of rapid climate oscillations evidenced in Greenland ice from the 2°C warmer climate of the previous interglacial era some 120,000 to 140,000 years ago (GRIP, 1993). This has led to the speculation that a world a few degrees warmer, from a few more decades of greenhouse gas buildups, might lead to a very nasty surprise: loss of current climatic stability.

WHAT IS HIGHLY SPECULATIVE?

Any prediction of what climatologists call the detailed regional distribution of climatic anomalies is speculative. It is still tough to predict confidently where and when it will be wetter and drier, how many floods might occur in the spring in California, or how many forest fires in Wyoming or Siberia will burn in August—although plausible scenarios can be given. How much sea level will change is also speculative (see, for example, Schneider, 1992), with most estimates ranging from a zero to one meter rise by 2100. Perhaps the best way to think about sea level rise is by a **probability distribution,** such as in Figure 2.3.

ECOLOGICAL IMPACTS: THE POTENTIALLY MOST SERIOUS CONSEQUENCE

Since the projection of regional climatic changes that evolve with time is still very speculative, so too is confident assessment of the agricultural, hydrological, ecological, or health consequences of global warming. However, we can construct a variety of plausible specific scenarios of climatic changes over space and time and then ask: "So what?" (Pearman, 1988; Smith and Tirpak, 1989). Indeed, such exercises have led to conflicting assessments of the agricultural consequences (NAS, 1991), and have increased concern for the hydrological consequences (Waggoner, 1990) and especially the ecological implications of most global warming scenarios (Peter and Lovejoy, 1992). Table 2.2 from a report by NAS (1991) illustrates this point. Let us examine the latter issue in more detail.

Although they include a wide range of uncertainty, the assumptions associated with the standard paradigm to assess the impact of global climate change are essentially surprise-free. Rather than postulating cases of low (or uncertain) probability in which little climate change or catastrophic surprises might occur, and then multiplying that low probability by the very large potential costs or benefits, most analysts are content to use a few standard **general circulation climate model** (GCM) CO_2-doubling scenarios to "bracket the uncertainty." However, I believe that a **strategic approach**—considering a very wide range of probabilities and consequences—is more appropriate (Figure 2.3), given the plausibility of surprises, even if we have but little capacity to anticipate specific details now. For example, two critical assumptions of the standard assessment paradigm are that climate extremes—drought, floods, super hurricanes, and so forth—either remain unchanged or will change with the mean change in climate according to unchanged variability distributions. How-

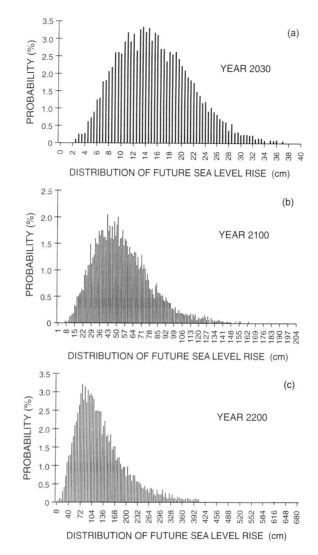

Figure 2.3

Plots showing the probability of various rises of sea level in the years 2030, 2100, and 2200, calculated on the basis of the "monte-carlo" estimation technique of Titus and Narayanan (1994).

ever, as Mearns, Katz, and Schneider (1984) have shown, changes in the daily temperature variance or the autocorrelation of daily weather extremes can either significantly reduce or dramatically enhance the vulnerability of agriculture, ecosystems, or other climate-extreme-sensitive components of the environment. How such variability measures might change as the climatic mean changes is as yet highly uncertain (Rind, Goldberg, and Ruedy, 1989; Mearns et al., 1990). Nevertheless, it could be that variability will change as climate changes, even though it is not possible, with current techniques, to ascertain credibly how that might occur.

Another assumption of the standard assessment paradigm is that "nature" is either constant or irrelevant in cost/benefit calculations (Daly, 1991). For example, "ecological services," whereby pest control or waste recycling are maintained, is assumed constant or of no value in most assessment calculations. Yet, as will be argued shortly, should climatic change occur in the middle to upper range of that typically projected, it is highly likely that communities of species will be disassembled, and significant alterations to accustomed patterns of pests and weeds seem virtually certain. Some people argue that pests, should their patterns be altered, can be simply controlled by pesticides and herbicides or medi-

Table 2.2

The sensitivity and adaptability of human activities and nature. (Based on data compiled from the "Adaptation" chapter of the National Academy of Sciences 1991 report, *Policy Implications of Greenhouse Warming*.)

	Sensitivity		
	Low	**Adaptation at some cost**	**Adaptation problematic**
Industry and Energy	X		
Health	X		
Farming		X	
Managed Forests and Grasslands		X	
Water Resources		X	
Tourism and Recreation		X	
Settlements and Coastal Structures		X	
Human Migration		X	
Political Tranquility		X	
Natural Landscapes			X
Marine Ecosystems			X

cines. The side effects of such controls are well known to many environment-development debates. However, what is not considered in the standard paradigm, but should be, is the consideration of a surprise scenario such as a change in public consciousness that would reject pesticide/herbicide application as an automatic procedure on the farm response to global changes. A related potential surprise is the recent suggestion that halogenated compounds typical of biocides are causing sinister damage to human and other animals' endocrine systems leading to animal and human health crises in the 21st century (for example, Colborn and Clement, 1992).

A NEW ASSESSMENT PARADIGM: INTERACTIONS AND SURPRISES

Although it has long been known that understanding and projection of the impact of human activities on climate and the converse involve detailed interactive studies of social, physical, biological, and chemical systems, the multidisciplinary nature of the analysis is exceedingly difficult to undertake in practice (Chen, 1981; Schneider, 1988). Several attempts to develop multidisciplinary assessments have spawned the International Geosphere-Biosphere Programme (IGBP, 1991) and the Global Change Program (Malone and Roederer, 1984). These national and international efforts certainly suggest a major recognition of the need to integrate physical, biological, and social factors. After such integration, we would be in a better position to make a quantitatively informed judgment on appropriate mitigation or adaptation policies.

Figure 2.4 is a simplified **system diagram** that stresses physical, biological, and social interactions as they might affect food security. I consider several examples at various points in this simple picture (simple only in the sense that it is much simpler than reality, although still very complex to analyze). These examples give an opportunity to identify plausible surprises and interactions. Indeed, interactions among physical, biological, and social systems may well prove to be the biggest surprise of all for adaptability of human systems to climate change. In our context, climatic influences on food security could, in turn, have major public health implications.

Figure 2.4 is my suggested economic/environment/food security assessment flow chart. It is designed to approximate the way one might model the food security system as affected by climate change, rather than fully describe reality. Although interconnected systems in reality do not flow linearly from the social sphere to food security as this flow chart suggests, if small enough "time steps" are taken, and the process is repeated many times, then the practical necessity of proceeding linearly should eventually result in only small errors.

The upper circle in Figure 2.4 represents, in essence, the "demand side" that leads to human activities responsible for documented global change forcings—greenhouse gas in-

Figure 2.4

System flow chart for diagnosing interactions between climatic change factors and food security.

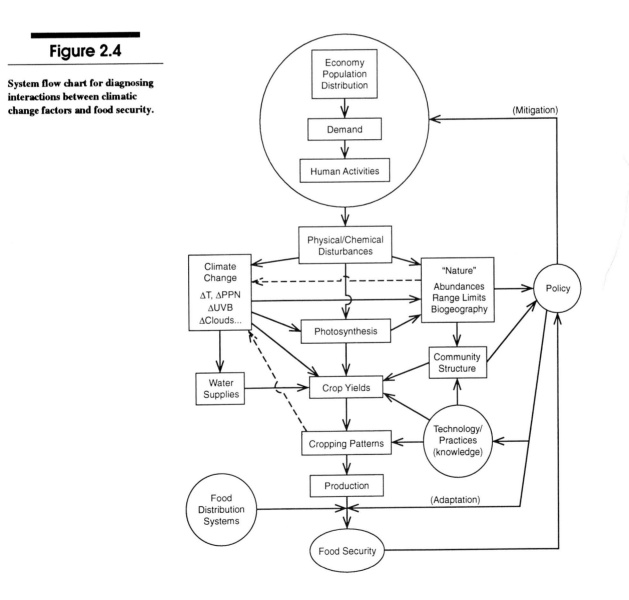

creases, increase in sulfur oxide emissions, habitat fragmentation, and so forth. Population, technology, and affluence are multiplicative factors that combine at global scale to provide scenarios for potential future physical and chemical disturbances. Based on these human behavioral assumptions, physical and chemical disturbances can then be supplied to the next level of analysis, such as models of climate change—typically in the form of general circulation models—to determine changes in temperature, precipitation, ultraviolet light (if ozone depletion is involved), a change in clouds, and so forth. In reality, an intermediate step is needed to determine how the natural and human disturbed carbon cycle operates, since the injected carbon dioxide and methane does not just sit in the system, but rather is actively transformed by atmospheric, oceanic, and biospheric chemical processes, as well as from direct disturbances to physical and biological sources and sinks (for example, deforestation or agriculture). Changes in decomposition rates of soil organic matter arising from feedbacks involving increases in temperature on increased soil microbial activity is a good candidate for "surprise," as is potential photosynthetic enhancement of crops, forests, and other biota. The latter has already been postulated by a wide range of people as a "positive surprise" (Idso, 1991; Rosenberg and Scott, 1994). That is why there is a direct arrow on Figure 2.4 from the disturbances to photosynthesis. Notice also a direct arrow from photosynthesis to "nature," since the effect of carbon dioxide enhancement of photosynthesis is not likely to force equal enhancements on all vegetative species (agricultural or natural).

It seems virtually certain that natural biological systems will primarily respond to CO_2 enhancement by conferring competitive advantage on some species at the expense of others. The phrase "winners and losers" has no traditional economic meaning in this context, for the conservation of nature is hard to define in terms of human value—if dollars are the sole measure of worth. The only statement that can be made with some assurance is that the larger the environmental system under stress and the more rapidly it is forced to change, the more communities of species will be altered by differential CO_2 photosynthetic response of species. This will likely lead to surprises, which currently are neither calculated nor valued, positively or negatively, in any current assessments of climate change, but rather are debated philosophically in the dichotomy of "economists versus environmentalists" (Myers and Simon, 1994).

It is already clear that changes in climate would change the abundances and range limits of many species (COHMAP Members, 1988; Davis, 1988, 1990; Root, 1992; McDonald and Brown, 1992). As Root and Schneider (1993) have noted, however, this combination of differential photosynthetic enhancement along with the differential response of various species to climatic changes will undoubtedly lead to difficulties in foreseeing reorganizations of biological communities. These have potential implications for evapotranspiration, which might in turn affect soil moisture and runoff, as well as implications for pest management, biodiversity preservation, and conservation of nature. For example, as Figure 2.5 shows, a change in climate predicted by these three standard **GCMs (general circulation models)** would dramatically alter the pattern of certain types of ticks in North America, substantially reducing the populations in hot southern zones, and dramatically increasing the population in northern zones, such as Nova Scotia (Smith and Tirpak, 1989). This has implications for human health as well as for natural communities, since insects are food for birds, which depend on localized vegetation for nesting, and which in turn serve as prey for animals such as foxes.

It is typically assumed that species may respond to climate change by shifting their range limit or abundances. However, most studies concentrate on individual species at the scale of a field plot the size of a tennis court (Kareiva and Andersen, 1988). How to "scale up" such studies to larger geographic and biological community levels is not yet based on a validated general theory, and has been treated in only a few studies (Levin, 1992; Root and Schneider, 1993; Ehleringer and Field, 1993). It is inferred from such studies that the migration rates

Figure 2.5

Simulated densities of ticks for selected cities under various scenarios of global warming resulting from projected doubled concentrations of atmospheric CO$_2$. The data plotted are based on results obtained by use of versions of the climatic general circulation models developed at the Oregon State University (OSU), the Geophysical Fluid Dynamics Laboratory (GFDL), and the Goddard Institute for Space Studies (GISS) and are compared with tick densities in the present climate (Smith and Tirpak, 1989).

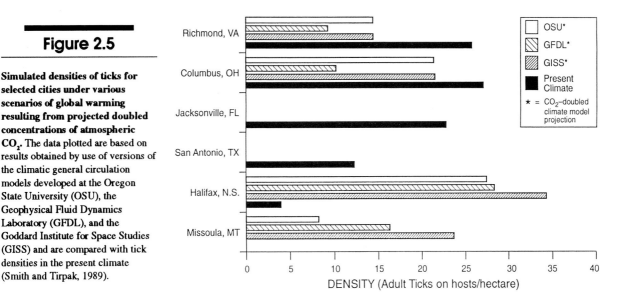

for specific species of vegetation, mammals, insects, reptiles, and birds are all different, yet the interactions among these species may have unique properties. This climate change could well disrupt the biocommunity structure.

The case of ticks suggests another category of surprises, for which there already has been considerable analysis: human health. Several studies have suggested that the direct effects of global heating on cardiovascular disease could lead to increased mortality rates for those who are vulnerable (Weihe and Mertens, 1991). Although it is usually noted that adaptations, ranging from simple acclimatization to proper choice of clothing, nutrition, diet, housing, and so forth, can substantially mitigate vulnerability to stress, there is potential for surprise impact if adaptations are not properly anticipated.

Referring back to Figure 2.4, the **food security** oval appears below the box labeled crop yields, which are affected by photosynthesis and water supplies. Water supplies, in turn, are affected by climate as well as by human adaptations such as fertilization and ground water pumping. Yields are also subject to influence from changes in biological community structure (the "nature" box on Figure 2.4) which includes pests, soil organic matter, and soil microbes. It is useful here to focus on climate-change/food-security impact assessment studies such as Rosenzweig and colleagues (1994). Recall that most such analyses of climatic impacts are based on the static forecast of an equilibrium CO$_2$-doubled GCM run, with the standard range of feedback uncertainties explicitly accounted for through choice of different GCMs. (We do not discuss here the consequences of adding a transient dynamic scenario of evolving climatic changes, and the tremendous potential increase in the surprise factor indicated by this additional dimension of analysis—see Schneider, 1994.)

Crop yields in Figure 2.4, then are translated to cropping patterns based on land use, technology, and knowledge which finally leads to overall production. Then, depending on economic and political factors that influence food distribution systems, food security can be ascertained. Currently, it is estimated that perhaps a half a billion to a billion people are "food insecure" due to dietary restrictions, political strife, poverty, maldistribution, and the normal variability of the weather resulting in droughts, floods, or weather-related pest attacks (Chen and Kates, 1994). Experts debate whether the world's doubled population will be food insecure by the mid-21st century (Daily and Ehrlich, 1992; Crosson and Anderson, 1992).

It is already a formidable scientific challenge to explain the range limits and abundances of most species today, even though they have adapted during thousands of years of very stable climate. Therefore, prediction of the transient response of biological communities faced with global climatic changes at 10 to 100 times faster rates than natural average rates of climatic change over the past 10,000 years is speculative at best! However, we do know some factors that can affect individual species and roughly how rapidly they can respond to various disturbances. Therefore, statements such as "disruptions of ecosystems," "tearing apart of communities of species," or even "ecological chaos" are plausible "forecasts" should global warming materialize at typically projected rates of 1°C to 5°C over the next 100 years.

Other highly speculative aspects of the global warming issue are the overall social and economic consequences of typical warming scenarios and the costs of actions to mitigate CO_2, CH_4, N_2O, or CFC emissions or whether to use technological schemes to offset warming (that is, so called geoengineering—see chapter 7 of this volume and, also, NAS, 1991). Should such costs slow development, this could also impact on human health.

Although state-of-the-art climatic models are far from fully verified for future simulations, seasonal and paleoclimatic (dealing with remote ages) validation exercises are strong evidence that modelers already have considerable forecast skill. Yet, uncertainties are as likely to render current best guesses underestimates as overestimates.

■

IS IT TOO EXPENSIVE TO ACT NOW?

The final, and perhaps most difficult, criticism made against those who propose actions to slow global warming is that immediate policy steps to cut CO_2 emissions are too expensive. For example, some newspaper ads by greenhouse critics suggest that if CO_2 emissions are cut the United States will be bankrupt and the Third World impoverished.

There is substantial Third World opposition to the prospect that they may not be able to have their own industrial revolutions as the developed countries did in the Victorian period when those then-developing Western countries used unregulated amounts of cheap and dirty coal to foster their industrial growth. Some now-developing Eastern countries, notably India and China, have abundant coal supplies. They would like to repeat Western history and use them as low-cost routes to industrialization. Of course, these countries in the 1990s have between them two billion people whereas the entire world didn't have two billion people in Victorian times. So the magnitude of the global impact of now-developing countries' use of coal to produce even half of the West's current industrial standard of living would be greater than that of developing Western nations in the past. Needless to say, such arguments are not greeted sympathetically in China or India, which correctly note the still large discrepancy in per capita consumption between developed and developing countries.

■

CAN CO_2 EMISSION RATES BE CUT 20 PERCENT?

One can reasonably argue that now-developing countries need not repeat the experience of Victorian industrialization with smog-choked cities, acid rain, and inefficient power production, given that modern technology has many better solutions. For example, electrical power generation efficiency today is near 50 percent, whereas it was half that at the turn of the century. Unfortunately, developing countries typically respond that high-technology, efficient power production is initially more expensive than the traditional options that are cheaper and more available to them. This dilemma sets up the obvious need for a bargain by which developed countries with technology and capital help to provide those resources to developing countries, which in turn develop their industries with the lowest polluting, most efficient technologies, even if they cost more cash up front.

Critics of emissions reductions cite the supposed annual cost of, say, a 20 percent CO_2 emissions reduction, at tens of billions of dollars annually. Some studies have suggested that carbon taxes to promote switching to less polluting energy systems could cost the United States "$800 billion, under optimistic scenarios of available fuel substitutes and increasing energy efficiency, to $3.6 trillion under pessimistic scenarios . . . to [the year] 2100." This quote from the February, 1990, "Economic Report of the President" to Congress was based on the initial results of the first wave of **economic model simulations.**

Let us pursue the economic modeling issue a bit further. When only the supposed costs to the economy (typically measured as gross national product—GNP—loss from a hypothetical carbon tax) are shown, seemingly staggering figures emerge. The costs, which result from "20 percent cut" carbon taxes imposed on one economic model used by Nordhaus (1992), are calculated to run into the trillions by 2105. The benefit of this "20 percent cut in CO_2 emissions," Nordhaus calculates, are a 30 percent reduction in global warming by 2105, only "worth" a percent or less of GNP loss by Nordhaus's assumptions—since he assumes the primary negative economic consequence of warming to be an agriculture/water supply loss of up to 1 percent of GNP in the United States (about $40 billion per year in 1990). In a later study, Nordhaus (1994) surveys the opinion of several dozen "impact assessment experts" who suggest the value to the global economy of averting typically projected global warming scenarios to 2100 run from a few percent of GNP up to tens of percent. Nordhaus (1992) puts no value on the potential for catastrophic effects on ecosystems or public health concerns or the security implications of long-term, very large climatic changes (perhaps up to 10°C in the 22nd century). Effects such as the tearing apart of communities of species or the impacts on the political stability of South Asia were tropical cyclone intensities to increase (a controversial debate) are not explicitly considered. These effects are implicitly included by Nordhaus's survey respondents (1994). But, to his credit, Nordhaus (1992) makes explicit the underlying economic growth assumptions that drive his conclusions. He assumes per capita consumption growing in all scenarios (his modest "optimal" scenario, the "uncontrolled" or business-as-usual scenario, and the so-called draconian scenario in which the "20 percent cut" carbon tax leads to the equivalent of a 20 percent emissions cut). In fact, there is about a 460 percent growth in per capita consumption from 1965 to 2105 in all his preferred scenarios, and even the draconian 20 percent cut carbon tax scenario still allows 450 percent world per capita consumption growth over this time period! In the value system of this analyst it is unconscionable to risk unabated, unprecedentedly rapid rates of climate change with potentially serious risks to public health, agricultural, hydrological, and (especially) ecological systems of the Earth merely because it will reduce per capita growth in human consumption from some 460 percent (the "uncontrolled" case) to "only" 450 percent (the draconian 20 percent cut case)!

In any event, although it is possible to use analytic methods and simulation models to investigate costs and benefits of different specific climate change or emission scenarios, these tools are even less well validated for long-term studies than climatic models and also do not include many significant factors the authors usually acknowledge (for example, see Oppenheimer, 1993; Dowlatabadi and Lave, 1993; Schneider, 1993; and Nordhaus, 1993). I believe all physical, biological, or economic analytic methods (sometimes known as **integrated assessments** when combined) are primarily useful to help a decision maker obtain a more complete knowledge of the potential benefits or risks of various alternative actions. This can aid him or her in formulating a subjective judgment that incorporates what is quantifiable with aspects that are not (for example, the value of a species facing extinction or the likelihood of altered disease vectors). It is a gross misunderstanding of analytic methods, be they climatic, ecological, economic, or integrated, to believe they provide the sole bases for arriving at "the policy answer" on either the effects, consequences, or mitigation aspects of global change.

■

"CLIMATIC INSURANCE"

The global warming debate then is both scientific and political. However, it is essential for the public to understand that there are vastly greater disagreements over what to do about the prospect of global warming (a political value issue) than over the probability (a scientific debate) that significant climate change is being built into the 21st-century climate. Estimates of climatic effects range from mildly beneficial (longer growing seasons) to highly catastrophic (more super-hurricanes or expanded disease patterns or mass extinctions). These uncertain impacts reflect the wide range of plausible climate futures forecast by most assessment groups, such as those seen in Figure 2.1. Although accelerated research will undoubtedly bring more reliable answers sooner, thereby helping to place decision making on a firmer factual basis, I do not believe it likely that any feasible level of research effort will pin down the detailed consequences of our continuing greenhouse gas emissions in less than a few decades—the time it will take the climate system itself to begin to answer the question.

Slowing down the buildup rate of the greenhouse gases that threaten unprecedented global warming does not require economically catastrophic measures, nor must it bankrupt industrial nations or doom poor countries to increasing poverty. Rather, prudent investments in energy-efficient equipment, houses, and power plants combined with sensible reforestation, population control materials for those who want them, and wasteful-consumption reduction programs, can both reduce the buildup rate of greenhouse gases and pay their own way. It is ludicrous simply to charge that greenhouse gas emission cuts are too costly without also weighing the economic, environmental, health, and strategic benefits of such investments. No individual or business ever got a return investment without making the investment first. That applies as well to societies and governments—the level at which global change problems occur and must be addressed.

Typically, impact assessments tend to focus on **best guesses** (for example, IPCC's best guess of 2° to 3°C warming by the middle of the 21st century), with uncertainty accounted for by selection of two or three equilibrium, CO_2-doubled model scenarios, or perhaps a low, middle, and high warming scenario. I certainly understand, even applaud, such bracketed analyses as important early steps in the attempt to incorporate uncertainty into physical, biological, and social components of environmental impact assessment. The possibility, however, of surprises that could accompany events that are not currently assessable or are of low probability suggests the obvious need for flexibility in managing systems that depend on environmental factors influenced by human activities, not least of which is climate change. I believe we should emphasize **strategic analysis** of a wide spectrum of probabilities and transient outcomes and assign to each of these scenarios some probability or range of probabilities—even though such probabilities will necessarily be subjective.

Every insurance policy has a premium, of course. But these environmental investments not only buy insurance against the possibility of unprecedented, possibly dire, climatic change but also can pay large dividends in the form of lowered energy costs, lower balance-of-trade payment deficits, less local air pollution, and diminished acid rain. In the case of public health, programs to improve services would mitigate the health impacts of potential climate changes and have other benefits as well. Therefore, such strategies make sense, even if global warming turns out to be as insignificant as some critics assert. In any case, since mild future climatic outcomes are as probable as catastrophic ones, and since it will likely take many decades to resolve major uncertainties, we can take the most cost-effective steps to lower risks now. Then, later, we can turn the pressure up or down as new assessments of the latest findings of the scientific community appear. Insurance for the planet is a logical extension of that safety net for individuals and firms. Unfortunately, the political will to cooperate internationally to mitigate potential global changes lags behind the need to extend insurance policies from families and firms to nations and the planet. The health of our species—and many others traveling on this planet—will depend on the outcome.

■
REFERENCES

Charlson, R.J., and Pilat, M.J. 1969. Climate: The influence of aerosols. *Journal of Applied Meteorology* 8: 1001–1002.

Charlson, R.J., Langner, J., Rodhe, H., Leovy, C.B., and Warren, S.G. 1991. Perturbation of the northern hemisphere radiative balance by backscattering from anthropogenic sulfate aerosols. *Tellus* 43ab: 152–163.

Chen, R.S. 1981. Interdisciplinary research and integration: The case of CO_2 and climate. *Climatic Change* 3(4): 429–448.

Chen, R.S., and Kates, R.W. 1994. World food policy: Prospects and trends. *Global Environmental Change* 19(2): 192–208.

Cooperative Holocene Mapping Project (COHMAP) Members. 1988. Climatic change of the last 18,000 years: Observations and model simulations. *Science* 241: 1043–1052.

Colborn, T., and Clement, C. (eds.). 1992. *Chemically Induced Alterations in Sexual and Functional Development: The Wildlife/Human Connection* (Princeton, NJ: Princeton Scientific Publishing).

Crosson, P., and Anderson, J.R. 1992. *Resources and Global Food Prospects, Supply and Demand for Cereals to 2030.* World Bank Technical Paper 184 (Washington, DC: World Bank).

Daily, G., and Ehrlich, P. 1992. Population sustainability and earth carrying capacity. *BioScience* 42: 761–771.

Daly, H.E. 1991. Ecological economics and sustainable development. In: C. Rossi and E. Tiezzi (eds.), *Ecological Physical Chemistry, Proceedings of an International Workshop* (Amsterdam: Elsevier), pp. 185–201.

Davis, M.B. 1988. Ecological systems and dynamics. In: *Toward an Understanding of Global Change* (Washington, DC: National Academy Press).

———. 1990. Climatic change and the survival of forest species. In: M. Woodwell (ed.), *The Earth in Transition: Patterns and Processes of Biotic Impoverishment* (Cambridge, UK: Cambridge University Press).

Dowlatabadi, H., and Lave, H. 1993. Letters. *Science* 259: 1381–1382.

Ehleringer, J.P., and Field, C.B. (eds.). 1993. *Scaling Physiological Processes: Leaf to Globe* (New York: Academic Press).

Graham, R.W., and Grimm, E.C. 1990. Effects of global climate change on the patterns of terrestrial biological communities. *Trends in Ecology and Evolution* 5(9): 289–292.

GRIP (Greenland Ice-Core Project) Members. 1993. Climate instability during the last interglacial period recorded in the GRIP ice core. *Nature* 364:203–207.

Hoffert, M.I., and Covey, C. 1992. Deriving global climate sensitivity from paleoclimate reconstructions. *Nature* 360: 573–576.

Idso, S.B. 1991. The aerial fertilization effect of CO_2 and its implications for global carbon cycling and maximum greenhouse warming. *Bulletin of the American Meteorological Society* 72(7): 962.

IGBP (International Geosphere-Biosphere Programme). 1991. *A Study of Global Change* (Stockholm: IGBP Press).

IPCC (Intergovernmental Panel on Climate Change). 1990. J.T. Houghton, G.J. Jenkins, and J.J. Ephraums (eds.), *Climate Change: The IPCC Assessment* (Cambridge, UK: Cambridge Univ. Press).

————. 1992. J.T. Houghton, G.J. Jenkins, and J.J. Ephraums (eds.), *Climate Change 1992: The Supplementary Report to the IPCC Scientific Assessment* (Cambridge, UK: Cambridge Univ. Press).

Jager, L. 1988. *Developing Policies for Responding to Climatic Change. A Summary of the Discussions and Recommendations of the Workshops Held in Villach, 28 September to 2 October, 1987.* WCIP-1, WMO/TD 225 (Geneva: World Meteorological Organization).

Kareiva, P., and Andersen, M. 1988. Spatial aspects of species interactions. In: A. Hastings (ed.), *Community Ecology: Workshop Held at Davis, California, April, 1986* (New York: Springer), pp. 35–50.

Karl, T.R., Kukla, G., Razuvayev, V.N., Changery, M., Quayle, R., Heim, R., Easterling, D., and Fu, C.B. 1992. Global warming: Evidence for asymmetric diurnal temperature change. *Geophysical Research Letters* 18: 2253–2256.

Kerr, R.A. 1992. Pollutant haze cools the greenhouse. *Science* 255: 682–683.

Levin, S.A. 1992. The problem of pattern and scale in ecology. *Ecology* 73: 1943–1967.

Lorius, C., Jouzel, J. Raynaud, D., Hansen, J., and Le Treut, H. 1990. The ice-core record: Climate sensitivity and future greenhouse warming. *Nature* 347: 139–145.

McDonald, K.A., and Brown, J.H. 1992. Using montane mammals to model extinctions due to global change. *Conservation Biology.* 6: 409–415.

Malone, T.F., and Roederer, J.G. (eds.). 1984. *Global Change* (New York: ICSU Press).

Mearns, L.O., Katz, R.W., and Schneider, S.H. 1984. Changes in the probabilities of extreme high temperature events with changes in global mean temperature. *Journal of Climate and Applied Meteorology* 23: 1601–1613.

Mearns, L.O., Schneider, S.H., Thompson, S.L., and McDaniel, L.R. 1990. Analysis of climate variability in general circulation models: Comparison with observations and changes in variability in $2xCO_2$ experiments. *Journal of Geophysical Research* 95: 20469–20490.

Morgan, M.G. 1993. Risk analysis and management. *Scientific American* (July), pp. 32–41.

Myers, N., and Simon, J.L. 1994. *Scarcity or Abundance: A Debate on the Environment* (New York: W.W. Norton).

NAS (National Academy of Sciences). 1991. *Policy Implications of Greenhouse Warming* (Washington, DC: National Academy Press).

Nordhaus, W. 1992. Rolling the "dice": An optimal transition path for controlling greenhouse gases. Unpublished paper presented at the annual meeting of the American Association for the Advancement of Science. Chicago, IL.

————. 1993. Letters—Response. *Science* 259: 1383–1384.

————. 1994. Expert opinion on climatic change. *American Scientist* 82: 45–51.

Oppenheimer, M. 1993. Letters. *Science* 259: 1382–1383.

OTA (Office of Technology Assessment). 1991. *Changing by Degrees. Steps to Reduce Greenhouse Gases* (Washington, DC: U.S. Government Printing Office).

Pearman, G.I. (ed.). 1988. *Greenhouse: Planning for Climate Change* (Melbourne, Australia: Commonwealth Science and Industrial Research Organization).

Peters, R., and Lovejoy, T. (eds.). 1992. *Global Warming and Biological Diversity* (New Haven, CT: Yale University Press).

Rind, D., Goldberg, R., and Ruedy, R. 1989. Change in climate variability in the 21st century. *Climatic Change* 14: 5–37.

Root, T.L. 1992. Effects of global climate change on North American birds and their communities. In: J. Kingsolver, P. Kareiva, and R. Huey (eds.), *Biotic Interactions and Global Change* (Sunderland, MA: Sinauer Associates), pp. 280–292.

Root, T.L., and Schneider, S.H. 1993. Can large-scale climatic models be linked with multiscale ecological studies? *Conservation Biology* 7: 256–270.

Rosenberg, N.J., and Scott, M.J. 1994. The implications of policies to prevent climate change for future food security. *Global Environmental Change* 4(1): 49–62.

Rosenzweig, C., Fischer, G., Frohberg, K., and Parry, M. 1994. Climate change and world food supply, demand and trade: Who benefits, who loses? *Global Environmental Change* 4(1): 7–23.

Schneider, S.H. 1971. A comment on "Climate: The influence of aerosols." *Journal of Applied Meteorology* 10: 840–841.

————. 1988. The whole earth dialogue. *Issues in Science and Technology* 4 (3): 93–99.

————. 1992. Will sea levels rise or fall? *Nature* 356: 11–12.

————. 1993. Pondering greenhouse policy. *Science* 259: 1381.

————. 1994. Detecting climatic change signals: Are there any fingerprints? *Science* 263: 341–347.

Schneider, S.H., and Mesirow, L.E. 1976. *The Genesis Strategy: Climate and Global Survival* (New York: Plenum).

SMIC (Study of Man's Impact on Climate). 1971. *Study of Man's Impact on Climate Report* (Cambridge, MA: MIT Press).

Smith, J.B., and Tirpak, D. (eds.). 1989. *The Potential Effects of Global Climate Change on the United States*. U.S. Environmental Protection Agency Document EPA-230-05-89-050 (Washington, DC: U.S. Government Printing Office).

Solow, A.R., and Broadus, J.M. 1989. On the detection of greenhouse warming. *Climatic Change* 15: 449–454.

Titus, J., and Narayanan, V. 1994. Probability distribution of future sea level rise. Unpublished manuscript.

Waggoner, P.E. (ed.). 1990. *Climate Change and U.S. Water Resources* (New York: John Wiley and Sons).

Weihe, W.H., and Mertens, R. 1991. Human well-being, diseases and climate. In: J. Jager and H.L. Ferguson (eds.), *Climate Change: Science, Impacts and Policy. Proceedings of the Second World Climate Conference* (Cambridge, UK: Cambridge Univ. Press), pp. 345–359.

Wigley, T.M.L., and Raper, S.C.B. 1991. Detection of the enhanced greenhouse effect on climate. In: J. Jager and H.L. Ferguson (eds.), *Climate Change: Science, Impacts and Policy. Proceedings of the Second World Climate Conference* (Cambridge, UK: Cambridge Univ. Press), pp. 231–242.

Woillard, G.M., and Mook, W.G. 1982. Carbon-14 dates at grande pile: Correlation of land and sea chronologies. *Science* 215: 159–161.

CHAPTER 3

ON THE SUSTAINABILITY OF RESOURCE USE: POPULATION AS A DYNAMIC FACTOR

■

John Harte*

■

INTRODUCTION

The current trajectory of human exploitation of natural resources is unsustainable. The reason why is simply stated: We are running down the environmental bank account by taking huge withdrawals from our capital reserves of clean air and water, fertile soil, biological diversity, and other resources. Ironically, we are at the same time degrading the capacity of natural ecosystems to regenerate or maintain those capital goods; that is, we are reducing nature's ability to provide clean air and water, to regulate flooding, to provide a tolerable climate, to maintain and regenerate fertile soil, and to provide habitat suitable for housing the Earth's living library of genetic information. Thus, at the same time that we are going into environmental debt through our massive withdrawals from the environmental bank account, we are hobbling the ecological processes that offer the only known means of maintaining the flow of deposits into the account.

To defend this view, we must look to where the devil is known to reside, which is in the details. In Part I of this chapter, I describe the dominant characteristic features and implications of our current trajectory of resource use. Rather than attempt a comprehensive review of natural resource exploitation, I focus on three important resources: energy, soil, and water.

The analysis in Part I is restricted to the endpoint of resource degradation and exhaustion, and the rates of resource use are simply assumed to equal population size times per-capita consumption. Thus, the magnitude of the human population plays only a passive role in Part I. It is people, of course, not economies or societies, that consume resources, but the role of people is reflected here only in the resource consumption figures; people are not treated as dynamic entities making choices about resource use or influencing the types and extent of the impacts of a given level of resource consumption. A related defect of the analysis in Part I is that it assumes the influence of population size is a simple one of proportionality. Thus, it neglects the existence of threshold phenomena in environmental science; by using resource consumption as a surrogate for environmental impact, some of

*Energy and Resources Group and Department of Soil Science, University of California, Berkeley, CA 94720

the most serious environmental consequences of human population growth are easily over-looked.

To remedy this, I examine in Part II the more dynamic and complex role that people—their numbers and their needs—actually do play in shaping both the trajectory of resource exploitation and its consequences. I argue here that the often-assumed notion that our woes are roughly linearly proportional to our numbers is outrageously simplistic and misleading in the sense that it understates the extent to which future generations are threatened by current population growth. In particular, it ignores a host of thresholds, synergism among multiple threats, and other nonlinear phenomena that tend to amplify risk and cause impacts of resource use to grow considerably faster than linearly in population size, even when the per-capita living standard remains constant.

Since I emphasize the environmental hazards of population growth here, my personal views on one issue need to be set forth clearly at the outset. The world's resources are unfairly distributed, with entire nations, and with large numbers of people within most nations, receiving far less than their per-capita share. There is no doubt that if resources were more fairly distributed, then the current size of the human population, and its rate of growth as well, would pose less of a challenge than they now do. Does this fact imply that we need not deal directly with population growth, and instead should focus all our effort on eliminating the inequities in resource distribution? I believe the answer is no. First, as Part I explains, even with perfectly equitable resource distribution worldwide, a growing human population will become increasingly resource limited, with all people equitably impoverished. Second, the extremes of democratic failure, dictatorship, and political chaos are more likely to flourish under conditions of scarcity and the poverty it induces, and either of these extremes breeds inequity and vitiates effective efforts to reduce it.

It is equally true that inequitable resource distribution, leading to an impoverished people that lack access to good health care, education, and job opportunities, often exacerbates population growth. Thus the connection between population growth and inequity is mutually reinforcing, and efforts to reduce population growth should be carried out hand in hand with efforts to eliminate inequity.

■

PART I: THE SUSTAINABILITY OF CURRENT PRACTICES OF RESOURCE USE

Here we look at the current trajectories of our consumption of three critical resources: energy, soil, and water. Our purpose is to sketch the essential reasons why, within a mere several generations, humanity will have to dramatically alter its current modes and trends of resource exploitation.

Energy

Consider, first, the expected lifetimes of the three principal **nonrenewable energy resources**—coal, petroleum, and natural gas—under three different sets of assumptions about the future rates of use of these primary nonrenewable fuels:

Scenario A. The future rate of fossil fuel consumption is fixed at 1990 levels, with any future growth in demand for energy met by nonfossil sources.

Scenario B. Future per-capita fossil fuel consumption is fixed at 1990 levels, so that future fossil fuel consumption grows at the average rate of population growth over the past 100 years, or roughly two percent per year.

Scenario C. Future per-capita fossil fuel consumption grows at the average rate it has been growing for the past 100 years; this, coupled with assumed population growth, implies that the total consumption rate of fossil fuels grows at four percent per year.

Scenario A is based on relatively optimistic assumptions about human behavior, although the technology does exist to do even better than this—that is, to reduce, not just stabilize, current fossil fuel use and still achieve a growing standard of living in the developing nations (Goldemberg et al., 1987). Scenario B is based on a more pessimistic but realistic assumption about population growth than is A. Scenario C is the scenario that best describes the historic trajectory of worldwide fossil fuel use over the past 100 years.

To convert these scenarios into estimates of resource lifetimes, we need to specify the sizes of the currently remaining recoverable (or **proved**) **fossil fuel resources** and the current rate of worldwide use of these fuels. These figures, along with the estimated lifetimes, are given in Table 3.1, which highlights the relatively brief duration of our future dependence on fossil fuels if population and per-capita-consumption growth rates continue at historical rates.

Turning next to **renewable energy sources**, we ask: For how long is dependence on solar-based renewable energy sources possible if historic growth rates continue? To explore this, we make the following optimistic assumptions about the exploitability of sunlight for direct conversion to electricity (for example, by photovoltaic devices) and the exploitability of biomass for producing chemical fuels:

1. Energy farms devoted solely to biomass for fuel can occupy 10 percent of the world's land area and biomass can be produced from sunlight at one percent efficiency.
2. Sunlight on five percent of the world's land area can be converted to electricity at 20 percent efficiency. Thermal energy currently used to produce electricity can be replaced with one-third that much energy content in the form of electricity from sunlight.

From these assumptions, it follows that dependence on solar-based renewable energy sources in scenarios B and C is possible for at most about eight and four generations, respectively (Table 3.2). Under scenario A (no growth in total demand), Table 3.2 indicates that sustainability is possible, but we emphasize that even scenario A may not be sustainable. In particular, the water resources needed to produce biomass harvests at the rate assumed may not be available. Moreover, continued extraction of biomass from land may gradually impoverish the soil, resulting in the need for fertilizers and other soil amendments at an unsustainable rate, which brings us to our next topic.

Lifetime of fossil fuels. Estimated lifetimes of coal, petroleum, and natural gas under three sets of assumptions about future consumption rates. Scenario A—rate of consumption of each fuel remains constant at current value; Scenario B—per-capita consumption remains constant but population grows at 2 percent per year; Scenario C—population and per-capita consumption each grow at 2 percent per year. Estimates of proved resources and consumption from WRI (1991).

Table 3.1

Energy Source	Estimated Proved Resource (10^{12} watt-years)	Current World Consumption (10^{12} watts)	Lifetime (Years)		
			Scenario A	Scenario B	Scenario C
Coal	750	3.4	220	84	57
Petroleum	170	4.3	40	29	24
Natural Gas	140	2.3	60	39	31

Table 3.2

Energy supply from renewables. Number of years to saturate renewable solar energy supply under the assumption that total world energy use (currently, 12.8×10^{12} watts) is supplied by direct solar conversion to electricity or by biomass. Scenario A—no growth in energy demand; Scenario B—two percent per year growth in energy demand; Scenario C—four percent per year growth in energy demand.

Source	Maximum Harnessable Rate of Energy Supply (10^{12} watts)	Years until Source Is Saturated		
		Scenario A	Scenario B	Scenario C
Solar Conversion to Electricity (5% of land area; 20% efficiency)	300	∞	158	79
Biomass Conversion to Chemical Fuels (10% of land area; 1% efficiency)	25	∞	34	17

Soil

About 10 percent of the world's land area, or 15 million square kilometers, is currently cultivated. Estimates of how much uncultivated land could be cultivated vary tremendously, depending on assumptions about the necessity of leaving land uncultivated, but a reasonable guess is that at most about another 10 million square kilometers is all that could be tilled without causing intolerable loss of natural ecosystems and of the goods and services those ecosystems currently provide.

To estimate how much additional time we buy by cultivating that additional land, assume that crop yields per hectare on currently cultivated land remain constant in the future and that those same yields can be achieved on the additional 10 million square kilometers of cultivable land (an assumption that is optimistic because the best land has generally already been put under the plow). Then if population growth is two percent per year, we can grow enough food to feed the human population at today's per-capita food production rate for only a bit more than one human generation, or 25 years.

Even if we make the absurd assumption that the entire land area of the planet can be cultivated and can yield on each new hectare today's harvests, so that we can expand tenfold current food production, then in a mere six generations we will have saturated the food-producing capacity of the land. Such is the inexorable mathematics of exponential growth.

Clearly then, unless we learn how to grow far more food on a hectare of land than we now can, the current trajectory of population growth is unsustainable. But, as we discuss next, the situation is even more grim than these numbers portray, and the reason is **soil loss and degradation**.

Loss of soil by erosion and degradation of in-place soil are of major concern for a number of reasons.

1. Loss and degradation affect the area of land on which people can productively grow food. Soil losses from agricultural land reduce the fertility of that land, resulting in both reduced crop yields and increased need for additions of energy-intensive industrial fertilizer.
2. Eroded sediments clog waterways and reservoirs, reducing the useful lifetime of reservoirs and necessitating dredging operations in bays and rivers. Increased risk of flooding can result as well from diminished channel capacity in rivers.

3. Eutrophication (overnourishment) of streams and lakes is due in part to the nutrients washed away within, or from, heavily fertilized, tilled soils.
4. Damage to coral reefs and other coastal ecosystems has been attributed to erosion from timber harvesting operations, road building on steep slopes, and tillage for crops (Darwin, 1851).

The current global rate of soil erosion has been estimated to be roughly 25 to 30 x 10^9 tons (soil) per year (Rozanov, Targulian, and Orlov, 1990). This is equivalent to the loss of about one millimeter per year of soil from all of the cultivated land on Earth, although the losses are not evenly distributed, of course, and not all of the eroded soil originates from cultivated land. This current rate of erosion should be contrasted with the estimated prehuman, or natural background, rate of loss of roughly 5 x 10^9 tons (soil) per year. In large areas of Africa and Asia, particularly China (Smil, 1984), the fertility of soil has been seriously reduced as a result of wind- and water-caused erosion. This is especially true in regions where demand for food precludes crop rotations and fallow periods, or where fragile soils that should never have been put to tillage are being exploited.

It is useful to consider the area of degraded soil as well as the tonnage of lost soil. Approximately 10 percent of the world's land area, or 15 million square kilometers, has been irreversibly lost during the past 10,000 years of human existence (an area coincidentally equal to that now cultivated). These losses occurred for a variety of reasons, but at the top of the list belong anthropogenic erosion, other anthropogenic transformations (such as deforestation) leading to desertification, the saltation of irrigated and poorly drained land, the submergence of land beneath artificial reservoirs, the paving of land, and land degradation due to natural climate change. The average rate of loss, over that period, was thus about 1,500 square kilometers per year.

According to one estimate, human activities are currently causing the loss of arable land at the rate of approximately 50,000 square kilometers per year, largely due to erosion and salination (Rozanov et al., 1990; see also WRI, 1993). At that rate, and assuming no growth in population, in 300 years we would lose all currently cultivated land.

If the rate of loss of arable land is assumed to be proportional to population size, and population growth continues into the future at a rate of 2 percent per year, then the 300-year estimate (above) drops to 97 years (Table 3.3). In other words, in about five generations, none of our current tillable land would be left. Putting under the plow the additional 10 million square kilometers of currently untilled land would only extend the lifetime of agriculture from 97 years to 120 years, not a major augmentation considering the ecological price that would have to be paid.

Table 3.3

Lifetime of arable soil. The estimates here assume a current rate of loss of 50,000 km^2 per year and that the loss rate is proportional to population.

Available Land		Lifetime (Years)	
		Fixed Population	Population Growth at 2% / Year
Current Tilled Land Area	15 x 10^6 km^2	300	97
Tillable Land Area	25 x 10^6 km^2	500	120

Water

Good water, like soil but unlike fossil fuel, is a resource for which there are no substitutes. Like the soil resource, natural processes play an important role in maintaining reliable supplies of freshwater. Water, like soil, can also be improved by human actions (at some expense) when it is degraded. Moreover, as is the case with soil, humanity must share water with natural ecosystems because of the numerous links between healthy natural ecosystems and the welfare of people. In other words, if we attempt to appropriate too much of a resource like water or soil, we risk losing more than we gain.

Worldwide, the average per-capita supply of total freshwater runoff in the late 1980s was roughly 9,500 m^3 per year (not including Antarctica); for the United States, the equivalent figure was roughly 8,300 m^3 per year (Shiklomanov, 1993). These numbers can be compared with the global and United States rates of water withdrawal for human use at that time, which were roughly 800 and 2,000 m^3 per year, respectively (Shiklomanov, 1993). In other words, we are appropriating about eight percent of available water worldwide, and 20 percent in the United States.

Falkenmark (1986) has estimated that a per-capita supply of approximately 1,000 m^3 of water per year is the absolute minumum required for drinking, sanitation, and other basic water needs in a moderately developed nation. In 1990, the potentially available water supply of 18 nations of the world was below this level. Most of the nations are in Africa and the Middle East, but the list also includes one from Central America (Barbados) and one in Europe (Malta). By the year 2025, a combination of population growth and water degradation will probably increase the number of nations lacking this basic supply to 33 (Table 3.4).

Unless we curb population growth, or find some environmentally and economically acceptable way to desalinate seawater, the number of nations below the critical level is likely to continue to grow larger still. When the world's human population is 9.5 times its current level, then the average per-capita supply will be below the critical level of 1,000 m^3 per year. At the current rate of world population growth, this will occur in about 120 years. Well before that happens, however, many nations will face an excruciating water crisis because of the unequal way in which water supply is distributed over the world and because of the expected rapid increase of water degradation with population growth (see Part II). As with the soil resource, our water supplies are not sustainable under current trends.

PART II: ENVIRONMENTAL IMPACTS AND THE SYNERGISTIC ROLE OF POPULATION SIZE

Many **synergistically destructive interactions** exist among the multiple stresses that we are inflicting on our planet (Harte, Torn, and Jensen, 1992; Harte, 1993). For example, deforestation generally exacerbates climate warming; climate warming exacerbates both the depletion of stratospheric ozone and the formation of tropospheric air pollution; and air pollution can contribute to forest loss. As shown below, population growth can intensify these destructive synergisms and trigger new types as well, causing impacts of resource use to grow considerably faster than linearly with population growth, even at a constant per-capita standard of living. The reasons have to do with the physical and biological characteristics of the environment and with facets of human nature as well.

To explore this, let us turn first to an energy-related problem—global warming and its impacts on water resources—as an illustrative case study. We then consider examples of soil degradation and other nonclimate-related aspects of water resource degradation to reinforce the major points. Finally, we discuss the unique role of feedback processes in environmental science, showing how they can readily lead to a dramatically disproportionate dependence of environmental impacts on population size.

Table 3.4

Sustainability of water use. Per-capita water availability for nations likely to face a severe water deficit by the year 2025 (from Gleick, 1993). Per-capita availability of 1,000 m³ per year is considered to be the minimum necessary for an adequate quality of life in a moderately developed nation (Falkenmark, 1986).

Continent	Nation	Water Availability (m³ per person per year)	
		Per Capita 1990	**Projected Per Capita 2025**
Africa	Algeria	750	380
	Burundi	660	280
	Cape Verde	500	220
	Comoros	2040	790
	Djibouti	750	270
	Egypt	1070	620
	Ethiopia	2360	980
	Kenya	590	190
	Lesotho	2220	930
	Libya	160	60
	Morocco	1200	680
	Nigeria	2660	1000
	Rwanda	880	350
	Somalia	1510	610
	South Africa	1420	790
	Tanzania	2780	900
	Tunisia	530	330
North and Central America	Barbados	170	170
	Haiti	1690	960
South America	Peru	1790	980
Asia/Middle East	Cyprus	1290	1000
	Iran	2080	960
	Israel	470	310
	Jordan	260	80
	Kuwait	< 10	< 10
	Lebanon	1600	960
	Oman	1330	470
	Qatar	50	20
	Saudi Arabia	160	50
	Singapore	220	190
	United Arab Emirates	190	110
	Yemen	240	80
Europe	Malta	80	80

Energy, Climate Change, Water, and Population

General circulation models (GCMs) indicate that climate change associated with increasing levels of greenhouse gases in the atmosphere is likely to lead to a drying of the soils during summer in the interiors of the major continents (Manabe and Wetherald, 1986). In response to a drying of cropland, people are likely to seek either new sources of irrigation water or new, wetter areas to farm. If desalination of sea water is the adopted solution, large amounts of energy will be required, which in turn will generate a host of additional ecological and economic problems. Impoundment of free-flowing rivers for additional reservoir capacity or importation of water from afar would degrade existing aquatic habitat. Converting new land areas to agriculture, assuming that new arable land exists, will nearly certainly degrade existing wetlands and terrestrial habitat.

No one, of course, could possibly predict today which combination of options will be selected by a future society in response to the threat of diminished food supply. These examples nevertheless provide a useful means of showing how environmental impacts of human activity depend in complex ways on human population size and why a doubling of population may lead to a far greater than twofold increase in the total impact of global warming. To explore this, it helps to break up the chain of impact into discrete steps, starting with fossil fuel consumption and concluding with the seeking of new water sources, our selected endpoint:

1. People consume fossil fuel.
2. Emissions of carbon dioxide to the atmosphere increase.
3. The atmospheric level of carbon dioxide rises.
4. The planet warms.
5. Soils dry.
6. Farm yields decline.
7. New sources of water are sought (potentially resulting in a host of impacts including those discussed above).

Let us look at these steps, in sequence, and assess how population size might affect the per-capita contribution to each.

1. While scenario builders often assume (as we did in Part I) that fossil fuel consumption will be proportional to population size (with the proportionality constant dependent on per-capita standard of living), in fact the so-called proportionality constant is likely to depend on population size as well as on per-capita standard of living. Alternative, nonfossil sources of energy—which, like biomass, need not contribute greenhouse gases to the atmosphere—might suffice for today's world of 5.4 billion people but not a world of 20 billion. Indeed, Tables 3.1 and 3.2 indicate that biomass could support at most roughly double the current population. As a result, if population were to stabilize, the average per-capita use of our fossil fuel resources would likely be less than if population continues to grow.

    In a more speculative vein, a world of 10 or 20 billion people might also be more fractious and politically unstable than a world of fewer people, so that nations will be unable to substitute new energy sources that are more abundant than biomass and more benign than coal. Of course, the worldwide per-capita standard of living might shrink with population size, causing per-capita use of energy in any form to shrink with population size. However, that is hardly a desirable way to negate a faster-than-linear dependence of fossil fuel use on population size, and ultimately it will also inhibit efforts to develop improved energy sources and to implement energy-conserving technologies.

2. Natural gas produces about three-fifths as much carbon dioxide per unit of derived energy as does coal, but the world's coal resource is an order of magnitude greater than

its gas resource. So again, if all else is equal, a world of 20 billion people reliant on fossil fuels will produce more carbon dioxide, per capita, than will a world of six billion people similarly reliant on fossil fuels. In other words, as population grows, the ratio of coal to natural gas consumption will likely increase and thus carbon dioxide emissions will grow faster than population size.

3. Under current oceanic and atmospheric conditions, the oceans are removing carbon dioxide from the atmosphere at roughly half the rate (two to three billion tons of carbon per year) that we emit it from fossil fuel burning. Thus, the rate of increase of atmospheric carbon dioxide will more than double if the emission rate doubles. In other words, the contribution at this step to the total per-capita impact is an increasing function of population size.

4. To examine the relation between the extent of buildup of greenhouse gases in the atmosphere and the extent of warming, let us take a brief look at some recent data from **Antarctic and Greenland ice cores**. These data shed some light on the question of whether climate warming is likely to be roughly proportional to the amount of additional greenhouse gases in the atmosphere or whether threshold phenomena exist that could cause disproportionately high responses to the buildup of these gases.

   Analysis of the Antarctic Vostok core has taught us that, over 160,000 years, temperatures varied in lockstep with the concentrations of atmospheric greenhouse gases such as carbon dioxide and methane. At first glance, this would appear to provide confirmation of our climate models that predict greenhouse-gas-induced warming trends over the next century; in fact, although the Vostok data do not contradict the well-established theory that greenhouse gases warm the climate, they do point to possible inadequacies in our current understanding of the full sensitivity of Earth's climate system. In particular, the Vostok data suggest that past warming trends caused significant increases in the amounts of greenhouse gases in the atmosphere. In other words, warming triggers positive feedback effects that further enhance the warming. Although we do not yet know what detailed combination of biological, physical, and chemical mechanisms trigger this positive feedback, we are forced to draw the conclusion that our current models, which ignore such feedback mechanisms, could be greatly underestimating the magnitude of warming that humanity is unleashing.

   Further evidence that the impending anthropogenic climate change may be more severe than our current models predict comes from ice core studies carried out in Greenland. These studies indicate that natural paleoclimate change during warming periods can be extremely rapid—with the speed of change as fast or faster than that predicted by current models of climate response to buildup of greenhouse gases in the atmosphere. The mechanisms causing these past rapid responses of the paleoclimate system are not well understood, which is one reason why they are not reflected in our models. Yet whatever the mechanisms, there is a significant likelihood that they will come into play as we warm the climate by the mechanism we do understand, namely, the direct radiative forcing from increasing greenhouse gas concentrations. Thus, we are again forced to the conclusion that our current models may be ignoring important mechanisms that amplify change.

   The ice-core evidence discussed above suggests the possibility of sudden increases in the rate and magnitude of warming when the forcing exceeds some threshold value. On the other hand, nature is on our side in at least one respect. The direct forcing effect of greenhouse gases on the planetary surface temperature increases more slowly than linearly in the atmospheric concentration of those gases. Perhaps at small increases in atmospheric concentration, the per-capita effect at this step, alone, is a decreasing function of population size while at higher atmospheric levels it increases with population size.

5. Evidence bearing on the dependence of evapotranspiration rates or of soil moisture on the magnitude of planetary warming is scant. Feedback mechanisms linking surface

temperature to soil moisture and hydrologic flows operate at a variety of spatial and temporal scales, and the combined effect is difficult to predict at present. The contribution at this step to the total per-capita impact could be an increasing or decreasing function of population size.

6. Consideration of plant physiology and soil physics indicates that crop yields fall faster than linearly in available soil moisture. As soils dry, hydraulic conductivity declines dramatically, leading to an increasing difficulty moving water from the root zone to the root itself. Moreover, the pressure potential of soil water (which directly influences the flow of water to the plant) decreases faster than linearly as soil water content declines and the plant wilting point is approached.

7. With increasing human population, the ability of people to meet their food needs with locally grown food declines; the historic response has been to seek larger, more distant sites for concentrated agricultural activity that can meet the needs of the growing population. One consequence of this is that the area of the region surrounding the agricultural areas, and from which water is potentially available, is proportionately smaller in comparison to water need. (Mathematically, the ratio of the perimeter of a circle to its area shrinks as the circle grows in area.) Thus, farmers are increasingly reliant on engineered water sources. Moreover, options for supplanting missing soil water (due to climate warming) by taking a relatively small fraction of water from peripheral areas at low environmental and direct monetary cost are reduced. The result is likely to be a growing reliance on the types of environmentally risky solutions discussed above.

To a first approximation, the total impact can be viewed as a product of these seven steps; in other words, impact equals (1) per-capita fossil fuel consumption, times (2) emissions per unit of fuel burned, times (3) increase in atmospheric carbon dioxide concentration per unit of emission, times (4) warming per unit of concentration, times (5) soil drying per unit of warming, times (6) decline in crop yield per unit of soil drying, times (7) likelihood of seeking environmentally disruptive new water sources per unit of crop lost. Because these individual factors are generally either increasing functions of population size (items 1, 2, 3, 6, 7), or the dependence is presently uncertain (items 4 and 5), it would be prudent to assume that the intensity of the specific impact endpoint analyzed is an upward-curving function of population size. Note that I am not attempting to predict the specific response of society to warming, in general, or to inadequate growth in water supply, in particular; rather, I am demonstrating that population might exert a dramatically nonlinear effect on the magnitude of the forces that will trigger some response.

The attentive reader will have noted that by considering the total impact to be a product of the separate factors, I have ignored the kind of complex interactions that I have demonstrated characterize the role of population; in particular, the magnitude of each of the seven individual terms in the product can depend on the magnitude of the others, and this could conceivably magnify even more the dependence of per-capita impact on population size.

Soil and Population

Soil loss can be exacerbated by a host of human activities that, unlike agriculture, are not directly dependent on the quality of tilled soil. Such activities are not tempered by the sort of direct feedback signals that farmers receive when their livelihood washes away in the streams or blows away in the wind. Erosion, for example, may be intensified by deforestation for fuelwood or for wood products, which leaves bare ground exposed to the force of raindrops. In the United States, the Forest Service has proposed clearcutting forested slopes in the western United States on the inadequately tested theory that this will reduce water loss by eliminating transpiration from trees. (See the collection of articles in *The Water*

Resources Bulletin, vol. 19, June 1983, for a discussion of this topic.) In the short term, this may indeed increase water yields in some watersheds. But over the long term, soil erosion from the deforested slopes will clog streams, rivers, and reservoirs, and thereby reduce the water supply for people. Those trees standing on the hillsides perform a valuable ecological service by holding the soil in place; the "price" the trees charge in transpired water is cheap indeed. And, besides, the water transpired from the western slope does fall as rain some-where else.

Poorly managed cattle grazing can also greatly increase soil erosion. This occurs because the hooves of cattle can destroy vegetation that holds soil in place, and because livestock often bite off plants so close to the ground that the plants do not grow back. Moreover, ranchers in many locations deliberately cut back the dense streamside vegetation (such as willow thickets along the banks of streams in the Rocky Mountains) to allow cattle access to water, thereby causing erosion of the stream banks. Finally, pressure to open more land to grazing is an important impetus to deforest tropical lands, particularly in Amazonia (Salati et al., 1990).

Increased soil erosion may also be triggered in the future if, as some have proposed (Goldemberg et al., 1987), we turn to crop residues for a major energy source. The reason is that soil to which organic residues are not returned can eventually lose its structural integrity and ability to resist the forces of wind and water.

We note that in each of the cases discussed above, population pressure is a major driving force that propels humanity to take actions that increase soil erosion. In a direct and often-discussed sense, this force is simply expressed in terms of more people requiring proportionately more wood products, meat, and energy, thereby leading to more grazing and deforestation and hence to more soil degradation. But the indirect forces are probably going to be at least as important in the future, as the combined constraints of water, energy, and other resource needs confront us. For example, more people require more water as well as more wood. Thus there is increased pressure on the U.S. Forest Service to augment water supplies by reducing transpiration "losses." More people require more energy as well as more food, and thus there is increased economic incentive to convert agricultural wastes, and even agricultural land, to energy supply.

Clearly, population does not enter the equation in a simple linear manner. Up to some population level, in the United States, there is no need to augment water supply in the West beyond the already highly managed system of reservoirs put in place over the past 100 years, or to turn to biomass on a major scale for an energy source. However, as population pressure rises above a critical point (that will depend on the physical environment, the specific resource or combination of resources needed, and the cultural, political, and technological characteristics of the society), the incentive to seek new "solutions" becomes hard to resist and new resource management practices are then set in motion. Their impact on soil erosion remains to be fully evaluated, but the information in hand suggests that dramatically increased rates of soil degradation can result.

Put differently, if each resource sector exerts pressure for exploitation merely in propor-tion to population size, then the pressure on society to implement resource practices that are mutually inconsistent and destructive can increase dramatically with population size raised to the power of the number of environmentally linked resources (like water and wood) on which society is mutually dependent. Worse yet, as the remaining resources decline in quantity and quality, then the pressure on society to provide more of each actually grows disproportionately faster than population growth, leading to a snowballing positive feed-back effect that can ultimately threaten the survival of a society as well as its resource base.

Worldwide, we are likely to witness over the coming decades an array of novel responses to human population pressure that will greatly alter the way people manage land, water, energy, and biological resources. The case of soil erosion illustrates how the choices we make to try to solve one problem can exacerbate others, leading to faster-than-linear growth of total environment impacts as population grows.

Water and Population

Again, as we did with erosion, it is useful to look at the complexity of the ways in which human numbers affect water supply, sketching some of the factors that contribute to the far-from-linear role population growth plays in creating scarcity.

At a low population density, a society may be able to derive its water from rivers, natural lakes, or from sustainable use of ground water. As population grows, so does the volume of water needed (we will assume demand is proportional to population size). Moreover, levels of waste discharge into the environment will grow as population rises. Thus, the available unmanaged supplies deteriorate at the same time that demand on them is increasing. So even at this stage, before the society attempts to augment supply, a **destructive synergy** is at work; population size affects the water resource in a manner that is not one of simple proportionality.

Consider, next, how societies actually do respond to the resulting water crunch. Traditionally they have impounded water, and this has led to increased evaporation rates. In arid regions, particularly, this can greatly diminish water supply. For example, at the lake impounded behind the Aswan Dam in Egypt, over 10 percent of the lake's capacity evaporates each year. Worldwide, evaporation from reservoirs amounts to nearly 200 km^3 per year (or a per capita rate of 40 m^3 per year), which is about seven percent of worldwide water consumption. For downstream water users, the effect of impoundment can be acute for another reason as well: Little or no water flows downstream from dams during times when reservoir managers are storing all available runoff.

Groundwater usage provides additional examples of this nonlinear population multiplier effect. Consider the case of an **aquifer** that currently supplies water to, say, one million people. Under the current climate and at current consumption rates, that supply might well be sustainable for the indefinite future. Now, suppose that the user population doubles (or per-capita water consumption doubles). At the existing recharge rate, the aquifer will then slowly draw down and perhaps will last for only 40 years, at which time the water will be too saline to drink. Note that in this example population doubling did not result in an 80-year aquifer lifetime becoming a 40-year lifetime but, rather, a sustainable aquifer becoming an unsustainable one. The impact on future generations far more than doubled.

The Oglala aquifer that underlies a large portion of the Great Plains in the United States exemplifies this phenomenon. The southern portion of this large underground reservoir has been drawn down because of overpumping for irrigation in Texas, resulting in the loss of about a quarter of its capacity in the past several decades (Postel, 1993). Before that, water extraction was believed to be roughly in balance with the recharge rate so that use of the aquifer was sustainable.

The problem of salt intrusion into groundwater supplies even more dramatically illustrates the enormous leverage that population growth can exert. When freshwater is drained from wetlands lying above groundwater supplies, such as has occurred in South Florida and Long Island, New York, to accommodate the space needs of growing populations, seawater intrudes into the aquifers (Harte and Socolow, 1971). Thus an aquifer that could sustainably provide water for a population of, say, one million people, can become saline and totally useless for human consumption in a matter of years, if the protective head of freshwater is removed from the land above because of even a small increment in population size.

Sea-level rise can also cause salt intrusion into fresh groundwater supplies, with only a modest rise capable of rendering unpotable an entire aquifer. This situation greatly concerns many island nations of the South Pacific that rely on a thin lens of freshwater beneath a relatively low-lying land surface. In particular, the sea-level rise expected to accompany global warming could reduce to zero the water supplies of many such nations.

The example of a wetlands providing protection to an underground water supply by preventing seawater from intruding into the aquifer is just one instance of a general pattern in nature: Natural processes tend to maintain a widely habitable environment, and when these processes are disrupted, the **carrying capacity** of the environment plummets. In effect, quality begets quality and degradation begets degradation. Reservoirs, created to provide adequate water supply when free-flowing rivers cannot do so, are often a breeding ground for disease vectors in developing nations as well as a source of evaporative losses. Pollutant levels can build up in reservoirs because of inadequate flow-through. Moreover, the water-cleansing capability of the biological communities in the sediments of natural streams and wetlands is lost when those water bodies are drained or confined in concrete channels. The population lever thus operates at a particularly insidious manner when population growth leads to ecosystem degradation—a doubling of population may lead to only a doubling of the area of ecosystem exploitation, or of some other measure of ecological resource degradation, but this may lead to a tenfold or more increase in the incidence of local human toxicity or disease.

Yet another way in which population size can exert a complex influence on water resources relates to the chemistry of the **acidification** process. Over large regions of North America and Eurasia the rain is polluted with acid. This pollution mainly comes from the burning of coal and oil, which leads to the emission of tens of millions of tons of sulfur dioxide and nitrogen oxides from smokestacks and tailpipes each year. These gases turn to acid—sulfuric and nitric—in the atmosphere and then fall to Earth in rain, snow, and as dry acid particles. Often the acids come back to Earth many hundreds of miles from where the gases that formed them were emitted. Hence, one nation's pollution can fall as acid in another nation. In some regions, these acids are so strongly concentrated in the rain and snow that they can increase the acid in lakes to levels that kill certain types of plants and animals. Populations of trout and salamanders in many lakes of eastern North America and Scandinavia have been wiped out as a result.

It is also possible that forests are being destroyed by acid rain. A phenomenon called **forest dieback**, characterized by the death of a large part of the forest over thousands of square miles, is occurring in West Germany. Forest dieback has spread during the 1980s and threatens the economic as well as the aesthetic value of large areas of Germany's revered forests. It is also occurring in Eastern Europe and possibly to a lesser extent in the eastern United States. Acid rain is suspected of contributing to forest dieback, but the actual combination of factors leading to forest dieback is not understood.

There are natural buffering processes that take place in most soils or waters and that provide some measure of protection against acidification. The acids in precipitation are neutralized by these processes and thus do no harm to organisms. The buffering processes, however, cannot continue indefinitely if the rate of acid input is too high. The reason is that the processes are carried out by chemicals in water and soil that can only be regenerated at some maximum rate; if they are used up in the buffering process at a rate faster than the rate at which they are regenerated, then eventually the buffering processes terminate. As a result, a lake can be subject for a long time to a low level of acidity in precipitation and exhibit no increase in its acidity. However, if the level of input exceeds some threshold, at some point (sooner if the level greatly exceeds the threshold) the buffering capacity of the lake water and the soils in the watershed are used up, and then the lake will rapidly become acidic. Thus, a doubling of the number of people consuming fossil fuel at a fixed per-capita rate can lead to considerably greater than double the level of acidity in downwind water bodies or soils.

The examples above indicate that water, like soil, can be degraded or lost at a rate that grows faster than linearly, with the rate of population increase. To conclude this section, I turn to a more generic issue, that of **feedback processes**, and show how they can lead to a highly nonlinear dependence of impacts on population size.

Feedback Processes and Population Size

When the global climate system, a specific ecosystem, an organism, or any other complex entity is disturbed by some perturbation, the effect on the entity (expressed, for example, as a temperature change, an alteration of species diversity, or a change in life span) is the sum of two terms. The first is the **direct response**. In the case of climate change from the buildup of greenhouse gases in the atmosphere, the direct response is the change in surface temperature due to the extra heat-absorbing capacity of these gases. This direct effect is roughly 1°C for a doubling of atmospheric carbon dioxide.

However, this direct warming can also trigger secondary or **indirect effects**. In the case of climate change, an example of an indirect effect is the change in the water vapor content of the atmosphere induced by the warming. Hot air holds more water vapor than cold air, so as the atmosphere warms, its water vapor content rises. This is called a feedback effect because water vapor is a greenhouse gas and thus the rise in water vapor content causes a further warming. Indeed, it is a *positive* feedback effect because the initial (direct) warming is enhanced by the process. Had warming led to a decrease in atmospheric water vapor, the feedback would be called *negative*.

The mathematical description of feedback is summarized in Figure 3.1. The **feedback gain**, denoted by g in the figure, is a number that is positive if the feedback is positive and negative if the feedback is negative. As shown, the total effect is the product of the direct effect times the factor $1/(1-g)$. When g is between 0 and 1, the factor $1/(1-g)$ is greater than one and so positive feedback causes the total effect to exceed the direct effect. (If g is equal to or greater than one, then feedback is so strong that the system goes unstable and the formula above no longer applies.) When g is negative, the factor $1/(1-g)$ is less than one and so negative feedback causes the total effect to be less than the direct effect.

Consider a situation where the magnitude of g is linearly proportional to population size. Then, as the figure shows, the total effect can grow considerably faster than linearly in population size. In particular, when g is near one, a very small increase in g can cause a huge increase in the factor $1/(1-g)$. Because the estimated value of g for global climate change is roughly 0.7, with a large range of uncertainty that might extend up to a value close to one (Lashof, 1989), this issue is not just academic.

Figure 3.1

A mathematical description of the feedback effect and population growth. As is discussed in the text, the feedback GAIN (g) is a number that is positive if the result of the feedback mechanism is positive, and negative if the feedback is negative. The TOTAL EFFECT is the product of the DIRECT EFFECT times the factor $1/(1-g)$.

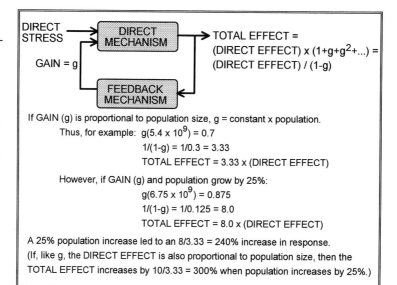

DIRECT STRESS → DIRECT MECHANISM →

GAIN = g

FEEDBACK MECHANISM

TOTAL EFFECT =
(DIRECT EFFECT) x $(1+g+g^2+...)$ =
(DIRECT EFFECT) / $(1-g)$

If GAIN (g) is proportional to population size, g = constant x population.
Thus, for example: $g(5.4 \times 10^9) = 0.7$
$1/(1-g) = 1/0.3 = 3.33$
TOTAL EFFECT = 3.33 x (DIRECT EFFECT)

However, if GAIN (g) and population grow by 25%:
$g(6.75 \times 10^9) = 0.875$
$1/(1-g) = 1/0.125 = 8.0$
TOTAL EFFECT = 8.0 x (DIRECT EFFECT)

A 25% population increase led to an 8/3.33 = 240% increase in response.
(If, like g, the DIRECT EFFECT is also proportional to population size, then the TOTAL EFFECT increases by 10/3.33 = 300% when population increases by 25%.)

In environmental science, are there a significant number of positive feedback processes in which the gain factor, g, grows at least as fast as linearly with population growth? It is certainly true that some very important feedback processes have gain factors that are essentially independent of population size. Consider, for example, the water vapor effect discussed above. The direct temperature change that triggers the feedback will depend on population (because people emit greenhouse gases), but the gain factor depends only on the physical properties of air and water.

On the other hand, there is at least one class of feedback mechanisms in which g does grow with population size—mechanisms that directly involve human response to the direct effect. For example, in a warmer climate, people may rely more on air conditioning, thereby burning more fossil fuel and augmenting emissions of carbon dioxide (see Figure 3.2). But the role of population need not be so blatant. Thus, carbon dioxide release from warmed soils may depend on population size because tilled and irrigated soils, particularly those with large amounts of stored organic matter, are particularly likely to undergo increases in soil decomposition rates in a warmer climate.

A systematic examination is needed of the major feedback mechanisms that have been identified in the environmental sciences. The purpose of that examination would be to determine the magnitude of the associated gain factors and the extent to which they depend on population size. More specifically, we need to know the extent to which the responses of people to environmental degradation will unleash a sequence of events that will exacerbate that degradation or ameliorate it. The natural sciences are not likely to provide all the needed answers because it is the behavior of people and their institutions that determine the extent to which gain factors depend on population size; insights from the social sciences and humanities are critical.

Figure 3.2

An example of a feedback mechanism in which the feedback gain (g) does grow with population size—a mechanism in which humans respond directly to the direct effect. As environmental temperature increases, people rely more on air conditioning thereby burning more fossil fuel, which augments emissions of carbon dioxide (CO_2), which further increases environmental temperature.

Summary of Part II

Space does not permit a comparable analysis of the influence of population on other resources or on all the other impact endpoints from global warming, let alone of all the endpoints from a wide variety of other environmental assaults ignored here, such as ozone depletion, overfishing, overharvesting of timber, pesticide use, smog formation, and so forth. However, having worked through similar exercises for at least some of these specific endpoints, it has become clear to me that we tend to greatly underestimate the threat that population growth poses to the well-being of our descendants if we assume our problems will at most increase in proportion to population size. The reason is that we tend to ignore the following three general principles that in various combinations apply to nearly all instances of resource use and environmental degradation:

1. The **"dose-response"** relation between environmental stresses (such as acidity of precipitation, or area of drained wetlands, or amount of carbon dioxide loaded to the atmosphere) and the response of the natural environment to that stress (such as acidification of soil and lake water, or useful lifetime of an aquifer, or amount of warming induced by carbon dioxide) are often nonlinear, with responses increasing faster than stress over a range of stresses relevant to current conditions. Thus, even if stresses are merely proportional to population, the responses increase faster than proportionally.
2. Numerous synergies exist among different kinds of environmental responses; these synergies are such that the impacts from deforestation, acidification, ozone depletion, climate change, erosion, water impoundment, pesticide use, and so forth tend to mutually reinforce each other, making the whole more like the product rather than the sum of the parts.
3. Human responses to resource depletion or environmental deterioration tend to trigger accelerated depletion and degradation in an intensifying spiral. Positive feedback processes in which the gain factor is proportional to population size generate impacts that can increase considerably faster than linearly with population.

The relation between population size and the sustainability of the human enterprise is complex indeed. For a specified means of achieving a constant per-capita level of well-being, impacts can grow far faster than linearly in population size. The general structure of that relation is such that considerably greater concern over population growth is warranted than has generally been shown by public policy makers. The types of synergistic interaction discussed above force us to reassess the traditional viewpoint that impacts are simply a product of population size times some population-independent measure of the impacts of achieving a given per-capita level of well-being.

■

CONCLUSION

We have seen that humanity is running down the environmental bank account of resources like clean water, air, and soil, and at the same time hobbling the ability of natural processes to replenish that account. We are simultaneously on a credit-card spending binge and a machine-smashing rampage within the factories that employ us. Out of the conviction as expressed by Harold Morowitz that "optimism is a moral imperative," however, I wish to conclude on a more hopeful note. The following is excerpted from the conclusion to *The Green Fuse: An Ecological Odyssey* (Harte, 1993).

Bold journeys characterize human history, from the prehistoric migration of peoples in response to the advance and retreat of the ice shelf to the great age of exploration of the new world. The most important migration of humanity may be just beginning, as we attempt to leave the dried and worn pastures of domination over nature and chart a new path that will take us to a sustainable and healthy future. It is a journey no less bold and challenging than those of the great explorers centuries ago.

Those early explorations would not have been taken by people hanging on to the old notion of a flat Earth over whose edge ships would fall; it took a new geography to give birth to a new, edgeless, vision of the human potential. Similarly, our forthcoming journey is stalled by another outmoded myth, one that concerns temporal rather than spatial bounds. Environmental science, my field of research, is particularly exciting now because it provides some hints of the next "new geography," a geography that might be parent to a timeless, sustainable, vision of humanity.

To think about all of yet unborn humanity, to consider its interests as we contemplate daily actions, is a tall order, however. Taken too literally, such notions could confound all decision making, but these abstractions actually do translate into razor-sharp arguments that can cut through a good deal of muddled thinking on practical and pressing issues. Here are a few examples:

When confronted with the frequently made argument that people are a good thing, so let's have more of them (and therefore ban contraception and scuttle family planning programs), we can reply, following Herman Daly: Yes, let's have as many people on this planet as possible—but not all at the same time. Let's populate the world with people for as long into the future as possible, by taking care that we limit population growth now.

To the argument that the banning of timber cutting in old-growth forests, oil drilling near marine sanctuaries, or strip mining in the wilderness will eliminate existing jobs, we can reply that if those natural resources are left intact, their inherent recreational and ecological value will create and sustain a far greater number of jobs for numerous future generations of workers. On the other hand, if these finite resources are mined greedily today, the ephemeral jobs will soon disappear along with the resources, permanently destroying future opportunities for a much greater number of jobs.

To the argument that we need to extract more rapidly our oil, gas, and coal resources to fuel growth of our industrial society, we can reply that the society that knows only how to burn fuel profligately will soon have none, whereas the society that knows how to use it efficiently can save barrel after barrel, generation after generation. Frugality and efficiency are the gifts that keep on giving.

To the argument that society faces a choice between the welfare of people and the survival of some obscure fish, flower, fungus, or frog species, we can reply: The well-being of people and the survival of species are synonymous. Sure, in the short run exterminating a wild species may confer on some people some advantage. However, like a quick pick-up from a narcotic, this advantage comes at the cost of placing the long-term health of humanity in jeopardy. The loss of every wild species is a loss of opportunity, both economic and aesthetic, for all the generations of our descendants, and the deliberate destruction of a species leaves a scar on the conscience of humanity that will never heal.

A sustainable future for humanity will only be achievable when people think and act in accordance with the reality that they are the future. As dreadful as is the current inequity in the distribution of resources between North and South, rich and poor, it pales in comparison with the impending inequity between us, living today, and those who will be born tomorrow and who, under current trends, are destined to inherit a rapidly deteriorating planetary life-support system.

My intent here is not to depress you with a litany of potential environmental disasters but rather to energize you with the suggestion that, at the deepest level, the solution to our environmental dilemmas is trivially simple. It requires no draconian hardships or new technological inventions, but merely a new way of thinking about ourselves and our relation to that most miraculous of all things on Earth—the future generations of life.

■

REFERENCES

Darwin, C. 1851. *The Structure and Distribution of Coral Reefs.* Facsimile edition (Berkeley: University of California Press).

Falkenmark, G. 1986. Fresh water—time for a modified approach. *Ambio* 15(4): 192–200.

Gleick, P.H. (ed.). 1993. *Water in Crisis: A Guide to the World's Fresh Water Resources* (Oxford, UK: Oxford University Press).

Goldemberg, J., Johansson, T., Reddy, A., and Williams, R. 1987. *Energy for a Sustainable World* (New Delhi: Wiley Eastern Ltd.).

Harte, J. 1993. *The Green Fuse: An Ecological Odyssey* (Berkeley: University of California Press).

Harte, J., and Socolow, R. 1971. *Patient Earth* (New York: Holt, Rinehart and Winston).

Harte, J., Torn, M., and Jensen, D. 1992. The nature and consequences of indirect linkages between climate change and biological diversity. In: R. Peters and T. Lovejoy (eds.), *Global Warming and Biological Diversity* (New Haven, CT: Yale University Press), pp. 325–343.

Lashof, D. 1989. The dynamic greenhouse: Feedback processes that may influence future concentrations of atmospheric trace gases in climatic changes. *Climatic Change* 14:213–242.

Manabe, S., and Wetherald, R. 1986. Reduction in summer soil wetness induced by an increase in atmospheric carbon dioxide. *Science* 232:626.

Postel, S. 1993. Water and agriculture. In: P. Gleick (ed.), *Water in Crisis: A Guide to the World's Fresh Water Resources* (Oxford, UK: Oxford University Press), pp. 56–66.

Rozanov, B., Targulian, V., and Orlov, D. 1990. Soils. In: B. Turner et al. (eds.), *The Earth as Transformed by Human Action* (Cambridge, UK: Cambridge University Press), pp. 203–214.

Salati, E., et al. 1990. Amazonia. In: B. Turner, et al. (eds.), *The Earth as Transformed by Human Action* (Cambridge, UK: Cambridge University Press), pp. 479–493.

Shiklomanov, I.A. 1993. World fresh water resources. In: P. Gleick (ed.), *Water in Crisis: A Guide to the World's Fresh Water Resources* (Oxford, UK: Oxford University Press), pp. 13–24.

Smil, V. 1984. *The Bad Earth* (London: Zed Press).

WRI (World Resources Institute). 1991. *World Resources 1990–91* (Washington, DC: World Resources Institute), p. 145.

———. 1993. *World Resources 1992–93* (Washington, DC: World Resources Institute), p. 112.

CHAPTER 4

BIODIVERSITY: WHERE HAVE ALL THE SPECIES GONE?

■

Mildred E. Mathias*

■

IMPORTANCE OF BIODIVERSITY

Biodiversity (biological diversity) is defined by Edward O. Wilson (1992) as "the variety of organisms considered at all levels, from genetic variants belonging to the same species through arrays of genera, families, and still higher taxonomic levels. Biodiversity includes the variety of ecosystems, which comprise both the communities of organisms within particular habitats and the physical conditions under which they live." Biodiversity is essential for the maintenance of ecosystem viability and function. We are learning more every day about the complex web of interrelationships that exists in the living world: the role of a large variety of animals in the pollination of flowers and in the dispersal of fruits and seeds; the importance of plants as chemical laboratories producing compounds that man has yet to synthesize; and the complexity of biotic and abiotic interactions. Every loss or modification of habitat affects biodiversity. However, in the present state of knowledge we cannot quantify these effects. Based on extrapolations of a few studies on discrete small areas, such as islets and isolated mountain tops, we can only make predictions.

We do not even know how many species exist. Approximately 1.4 million have been described (Figure 4.1), but more and more are being found as we explore more of our world. Predictions of the total number range from a conservative estimate of 12.5 million species to possibly as many as 100 million, mostly insects and microorganisms, enormous groups about which we still know very little. As human populations increase, the modification and sometimes total loss of habitats may result in the extinction of species faster than we find new ones and certainly more rapidly than we learn of their function in the ecosystem in which they occur (Erwin, 1991; May, 1988, 1992; Wilson, 1992).

Diverse flora and fauna have developed and gone extinct many times in the history of the Earth. They are known to us only from meager fossil records. Tectonic movements have disrupted entire ecosystems with fragmentation and isolation. There have been various

---

*Department of Biology, University of California, Los Angeles, CA 90024-1606, Deceased, February 1995.

Figure 4.1

About 1,400,000 species of organisms are known to science—about three-quarters of which are animals, and more than half of which are insects—but this represents only a fraction, probably less than 10 percent, of all species that currently inhabit our planet. (Modified after Wilson, 1992, p. 134. Reprinted by permission of the publishers from *The Diversity of Life* by Edward O. Wilson, Cambridge, Mass.: The Belknap Press of Harvard University Press, Copyright © 1992 by Edward O. Wilson.)

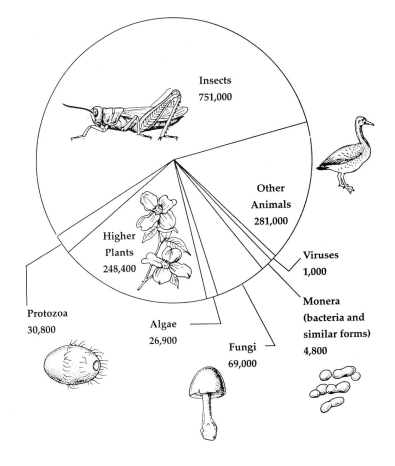

ALL ORGANISMS: ~1,400,000 Species

estimates of the **background extinction** rate based on our knowledge of the fossil record. The rate has not been uniform, and major extinctions have occurred in a number of relatively short time periods during the past 600 million years. Extinction rates between the episodes have been relatively low. Groombridge (1992), assuming that the average lifetime of a species in the fossil record is about four million years, estimates an extinction rate of four species each year out of a total number of species of around 10 million. Wilson (1992) assumes that past species lived about one million years and the normal background rate of extinction is about one species per year per one million species. Extinction rates vary for different groups of organisms. The average duration of a vertebrate species is estimated to be about five million years, and 900,000 have become extinct every one million years during the past 200 million years. Thus, the background rate for vertebrates is approximately 90 species each century (McNeeley et al., 1990). Myers's (1988) estimate for higher plants is that one species has become extinct every 27 years over the past 400 million years. Wilson (1992) has mentioned the paradox of biological diversity: "Almost all the species that have ever lived are extinct, and yet more are alive today than at any time in the past."

There have been some recent catastrophic extinctions, but none on the scale of the present extinction caused by human activity. The devastation caused by the eruption of Krakatau in 1983 has been described by Wilson (1992). A remaining piece of the original volcano, the island of Rakata, is reforested, but with a disharmonic vegetation (Mathias, personal observation in 1971). The eruption of Mt. St. Helens leveled all vegetation in a large area, and it is being repopulated slowly. Similarly, the lava flows of Kilauea on the island of Hawaii and Volcán Arenal in Costa Rica have eliminated many habitats (Del Moral and Wood, 1993; Larson 1993; Lipman and Mullineaux, 1981; Wilson, 1992). These natural catastrophes have destroyed or modified entire ecosystems and almost certainly caused the extinction of species.

Intense hurricanes, Andrew in south Florida and Iniki on the island of Kauai, Hawaii, have destroyed native habitats as well as individual species. On Kauai, where only 14 individual plants were known of *Cyanea acerifolia* (Lobeliaceae), all were destroyed in their native habitat; only two of the four known individuals of the palm *Pritchardia viscosa* are left in the wild. However, the major damage to the habitats from these hurricanes has been the immediate invasion of exotic weeds that will delay or inhibit regeneration of the native vegetation and repopulation of the original fauna. The entire ecosystem has become less diverse (Mathias, personal observation; Ogden, 1992; Perlman, personal communication).

EARLY HUMAN-CAUSED EXTINCTIONS

Very early man probably had little effect on the world in which he lived. However, we have some data for those times when populations increased and major movements of humans occurred around the world. With the arrival of the first humans on Australia about 50,000 years ago, nearly all the species of large mammals and large flightless birds began to disappear. Similarly, in the Western Hemisphere, 73 to 80 percent of the large mammals disappeared.

These losses of the largest animal types were repeated again and again—the dodo of Mauritius, the giant elephant birds of Madagascar, the turtles of the Galapagos. Between 1838 and 1888 it is estimated that over 13,000 giant tortoises were removed from the Galapagos to be used as food for ships' crews (Robinson and Redford, 1991). Changes in biodiversity due to **subsistence hunters** continue to this day. Robinson and Redford (1991) have reported that in three Waorani villages in Ecuador, hunters killed 3,165 mammals, birds, and reptiles in less than a year. These included 562 woolly monkeys, 342 Cuvier toucans, and 152 white-lipped peccary. In the state of Amazonas, Brazil, with a rural population of 575,000, the average annual kill for food was 2.8 million mammals and 530,000 sea birds, as well as 200,000 reptiles, for a total of about 3.5 million vertebrates. In 1959, 6,500 manatee were killed, and although they are now protected, illegal manatee are still sold in local markets. In the Tocantín River drainage of Brazil the annual trade in caiman is 21,500 to 32,000 individuals. Animal elimination in these numbers almost certainly affects the local ecosystem. Groombridge (1992) has summarized animal extinctions in 30-year intervals from 1600 to 1959, with only three verified animal extinctions before 1629, only 25 by 1749, with a marked increase beginning in 1750, and a total from that date to 1959 of 322 animal species. Before the arrival of Europeans, almost all of Central America was forested. By 1985 less than 40 percent of the forest remained in the seven countries (Groombridge, 1992). Most of the forest clearing has been for agriculture and cattle grazing.

CURRENT RATES OF EXTINCTION

Wilson (1992) provides a conservative estimate of the current extinction rate of 1,000 species a year from destruction of forests and other key habitats in the tropics. In this decade he estimates that extinctions may rise past 10,000 a year and by 2020 we may have lost a fifth or more of all extant plant and animal species. This vastly exceeds rates in recent geological and human history and is enormously greater than the development of new species by ongoing evolution. He states that large animals especially, such as condors, rhinoceros, manatee, and gorillas, are close to extinction.

There are a few well-documented records of the precise time of extinction of an entire species. Matthiessen (1959) has provided one, an account of the extinction of the great auk, a bird that was once common on the ocean rocks of northern Europe, Iceland, Greenland, and the maritime provinces of Canada. The birds were slaughtered indiscriminately for flesh, feathers, and oil. In June 1844, Icelanders killed the last known auk, and none has been seen alive since.

David Norton (1991) has written an "obituary" for a plant, an endemic mistletoe, *Trilepidea adamsii*, parasitic on common understory shrubs and small trees, with a limited natural distribution on North Island, New Zealand. Despite searches at known localities the species has not been seen since 1954 and is not known in cultivation. Deforestation reduced its habitat and, as bird populations dropped for the same reason, seed dispersal was reduced. Plant collectors took specimens for unsuccessful cultivation and scientific study. The introduced brush-tailed opossum may have browsed the last plant.

Varley (1979) has given us the story of another endemic plant, *Trochetia erythroxylon*, a shrub on St. Helena and fortunately still in cultivation. In May 1976, only two shrubs remained on the island, one high up on an exposed position in one of the small remnants of the original forest, where it was battered by the winds and apparently no longer flowering. The other plant was in cultivation in a more sheltered position outside the forestry office. This plant flowered and set seed each year and the seed readily germinated. Seedling plants were given to local amateur gardeners, but the plants usually died. Soil samples from the original forest site were analyzed and proved to be very acid (pH 4.1) with a 50 percent content of organic matter. Rainfall at the site was 1,200 mm a year. Fresh seed from the cultivated plant was taken to the Royal Botanic Gardens, Kew, where it was germinated, planted in synthetic soil, fed with acid chemicals and distilled water, and the plants flourished. This species could no longer survive in its native habitat where the forests had been cut over centuries and goats were introduced for meat. The goats ate any seedlings, preventing regeneration of the forest. Through time the acid top soil eroded away. The rocks underlying the acid soil were alkaline (pH 7.8) and, consequently, the new soil was alkaline. Uplift of air over the barren eroded areas resulted in a decrease in rainfall that then averaged only 530 mm a year where the plant was once abundant, a profound modification in habitat due to man's activities.

Soulé and Kohn (1989) have described the extinction of the large blue butterfly in Great Britain where it lived in open areas that it required because its caterpillars developed in the nest of an ant that could not survive in overgrown vegetation. The reduction in grazing, a human-mediated activity, resulted in the return of the original closed vegetation in which neither the butterfly nor the ant could survive.

The world's last dusky seaside sparrow died in an enclosure at Disney World in 1987 where attempts were being made to breed it with another subspecies. Terborgh (1992) has given us an excellent account of the problem of the steep decline of the North American migratory songbirds. It is not simply due to destruction of their wintering habitats in the neotropics, although that is a contributing factor. Grassland nesting birds return to find their summer habitats converted to agricultural fields treated with chemical fertilizers and herbicides; forest fragmentation has led to an increase in nest predators; and water birds, such as the ducks, return to drained prairie wetlands with available water declining in quality.

These few examples of decline or extinction demonstrate that the causes are not always due to a single activity, but often a much more complex set of interactions.

■

POPULATION EXTINCTION AND GENETIC DIVERSITY

Populations are the fundamental evolutionary units. A species may consist of a single population, perhaps with only a single individual, as in the case of the one remaining tree of *Trochetia* on St. Helena, or it may have billions of individuals in many populations. The **population** is a pool of genetic variability that acts as an evolutionary unit through time. The loss of this variability may result in reduction in the ability of populations to survive in changing environments. Fragmentation of populations resulting from habitat modification is of major concern today. Small and isolated populations often lose variability due to genetic drift and are more subject to extinction (Barrett and Kohn, 1991; Ehrlich and Murphy, 1987).

A few examples of differences in species and in populations are of interest. Torrey pines (*Pinus torreyana*), for example, have no detectable genetic variation within the two known localized populations. In contrast, the knobcone pine (*Pinus attenuata*), which is widespread, though fragmented into a number of populations, showed more genetic variability than is normal for conifers (Ledig, 1986). Huenneke (1991), summarizing the results of several studies on the genus *Eucalyptus*, reports that localized species have little genetic diversity, in contrast to widespread species that have greater genetic variability. Wilson (1992) has written a masterful and readable account of *The Forces of Evolution*. Falk and Holzinger (1991) have collected a number of papers on genetic diversity in plants and provide an extensive bibliography.

■

HUMAN MEGA-DESTRUCTION

The mega-destruction that humans are inflicting on the Earth today is at a scale not seen before: thousands of hectares of forest clear-cut, burned, and converted to agriculture; other thousands going under concrete and blacktop; rivers channelized and dammed; estuaries drained for marinas or filled for development; freshwater ponds and vernal pools drained and laser-leveled. These activities seal the fate of populations, species, and often entire ecosystems.

In an effort to prioritize conservation activities, 18 areas of the world have been identified (Myers, 1988, 1990) as "**hot spots**," ecosystems with many **endemic species** (found nowhere else) in greatest danger from human activities (Figure 4.2). They are areas with an immediate need for inventories and planning. The designation of hot spots is a shift from looking only at the rare and endangered or threatened species, the beautiful terrestrial orchid, or the charismatic panda, cheetah, or parrot (McNeely et al., 1990; Myers, 1988, 1990; Wilson, 1992). Several of these areas are in Mediterranean climates, favorable for urban development.

One hot spot is the coast of **Chile**, with over half the Chilean flora crowded into six percent of the country, the most densely populated area. It is estimated that only one-third of original flora has survived (Takhtajan, 1986). A second area is the **Cape Floristic Province** of South Africa, where, in 89,000 square kilometers still surviving, there are 8,000 plant species, 73 percent endemic in the specialized fynbos (winter wet and summer dry vegetation similar to the North American chaparral). Agriculture, development, and the introduction of exotic plant species have already taken a third of the area, and much is degraded. At least 26 species are extinct, and 1,500 are rare or threatened. The Cape Region is a small portion of

Figure 4.2

"Hot-spot habitats," regions in which ecosystems occur that contain many kinds of plants and animals ("endemic species") that are found nowhere else and that are in the greatest danger from human activities. (Modified after Wilson, 1992, pp. 262–263. Reprinted by permission of the publishers from *The Diversity of Life* by Edward O. Wilson, Cambridge, Mass.: The Belknap Press of Harvard University Press, Copyright © 1992 by Edward O. Wilson.)

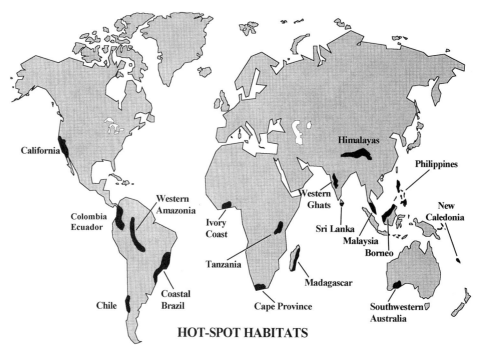

HOT-SPOT HABITATS

southern Africa and, if we include the rest of South Africa, Lesotho, Swaziland, Namibia, and Botswana, plant endemism is 80 percent of the total flora of about 23,200 species (Takhtajan, 1986).

Another hot spot, **southwestern Australia,** has over 3,600 species, 78 percent found nowhere else, and one-fourth are rare or threatened. In recent years agriculture, mining, exotic weeds, and wildfires have taken their toll. The southwest is a small part of Australia that has a high percentage of endemics, with seven families of mammals, four of birds, and 12 families of flowering plants—far more endemic families than any other country. At the species level for vertebrates and higher plants, endemism is 81 percent. Ninety-five species of vertebrates are listed as threatened, and 16 species of mammals are believed to have gone extinct due to human settlement and invasions by introduced species (Takhtajan, 1986).

The fourth Mediterranean hot spot is the **California Floristic Province,** extending from southern Oregon to Baja California, Mexico, and containing one-fourth of all the plant species found in the United States and Canada combined. The present flora has been evolving slowly through time, but it is now being seriously impacted by humans. Where once in the past there were magnificent wildflower fields in the southern San Joaquin Valley, now there are orchards, vineyards, cotton fields, and expanding cities. Oak woodlands are invaded by suburbia. Golf courses multiply, and even the dry washes are threatened. We can look at the extensive list of rare and endangered native plants updated regularly by the California Native Plant Society to find the causes of the threats—all directly or indirectly related to human activity. In addition to the massive destruction of plant populations for urbanization, agriculture, and grazing, the causes include the effects of stream channelization and dams, mining, road construction and maintenance, feral animals and exotic plant introductions, off-road vehicles, commercial and private plant collectors, wildfires, herbicides, foot traffic, displaced native animals, ground water pumping, hydroelectric developments, dumps, waste water disposal, fluctuating water levels, lowered water tables, recreational use, timbering, and clear-cutting. A number of endangered plants are known from only one location and a few from single individuals. Among those that have not been seen for at least 50 years, even though there have been extensive searches, are *Sibara filifolia,* last recorded on Santa Cruz

Island, California, and *Phacelia cinerea*, last seen in 1901 on San Nicolas Island, both probably destroyed by feral animals. *Potentilla multijuga* was last collected at Ballona, Los Angeles, in 1890. *Scheuchzeria palustris* var. *americana,* collected last in 1897, was probably buried under Lake Almanor.

With increased pressures on fresh water for humans and agriculture in California, some of the most threatened habitats are in aquatic and wetland ecosystems. In 1850 there were approximately two and a half million hectares of permanent, seasonal, and tideland wetlands. It is estimated that over 90 percent of these wetlands are gone, converted to agriculture, to urban development, or dammed and channelized for irrigation and flood control. Levees and reclamation have permanently changed many ecosystems. Agricultural practices typically affect wetlands directly on site with installation of drainage systems, levees, and irrigation delivery systems. The indirect effects on the aquatic and wetland ecosystems include residues from fertilizers and pesticides. It has been estimated that 35 percent of the rare and endangered animal species are in some way dependent on wetlands. Large numbers of bird species require wetlands for nesting, feeding, resting, and wintering habitats. In California 65 percent of the 113 native fish are either extinct, officially listed as in danger of extinction, or need special management to protect them. Coastal wetlands are important spawning and nursery areas for over one-half of the commercial marine fishes. Vernal pools in California contain some of the rarest plants and animals. There have been many populations of vernal pool organisms lost through conversion to agriculture. Riparian forests provide critical corridors for animal and plant dispersion in wildland watersheds and downstream river systems. It is essential to develop integrated management of our remaining aquatic systems and wetlands if loss of biodiversity is to be reversed (Frayer, Peters, and Pywell, 1989; Mathias and Moyle, 1992).

The remaining hot spots are in the wet tropics and subtropics where diversity is high and inroads by man are increasing. Tropical forests are vital in the maintenance of biodiversity, and they are the most biologically diverse communities and the most complex ecosystems known (Lovejoy, 1992). They cover only about seven percent of the land area of the Earth, yet estimates of the percentage of the world's plant species in them range from 50 to 90 percent. These forests are far from uniform in their plant communities; variations over short distances depend on soils, hydrography, exposure, wind, and certainly other abiotic factors. In a half-day walk in a lowland tropical wet forest one can pass from lowland seasonally flooded forest surrounding an oxbow lake, or from a river with cane brakes and lowland trees such as figs, cecropias, and coral trees, among others, into an ecotone of bamboo thickets, and on into drier well-drained forest that supports the major timber trees. Palms form thickets along the edges of oxbow lakes and in time may completely take over the wet area. There may be clusters of a single species and also many isolated individuals of others. Unless large areas are saved, many unique communities may be lost (Lovejoy, 1992; Mathias, personal observation).

The seasonally dry or monsoon forests of the tropics were the first to suffer the inroads of man. They were easier to cut in the long dry-season, and they provided a somewhat more desirable area for settlement than the extremely moist forests of such places as the Amazon Basin or the wet forests of eastern Madagascar. In western Ecuador, more than 90 percent of the dry forests disappeared between 1938 and 1988 (Figure 4.3). Similar losses have occurred from southern Mexico throughout Central America (Figure 4.4) and Panama (Janzen, 1986, 1988). Ten of the hot spots are in the tropical moist or wet forests, popularly referred to as rainforest.

With continued timbering in most parts of the world, fragmentation is occurring at a rapid rate, resulting in forests that are too small to remain viable. The tropical hot spots include the eastern forests of **Tanzania,** isolated from human influence until the last century, rich in endemic species of African violets, the parents of the common garden impatiens, and wild coffee species, and now suffering increasing pressures from exploding populations with concomitant agricultural development and logging (Wilson, 1992).

Figure 4.3

Deforestation of Ecuador has been severe over the past half-century: From 1938 to 1988, more than 90 percent of the forests of western Ecuador have been destroyed. (Modified after Wilson, 1992, p. 265. Reprinted by permission of the publishers from *The Diversity of Life* by Edward O. Wilson, Cambridge, Mass.: The Belknap Press of Harvard University Press, Copyright © 1992 by Edward O. Wilson.)

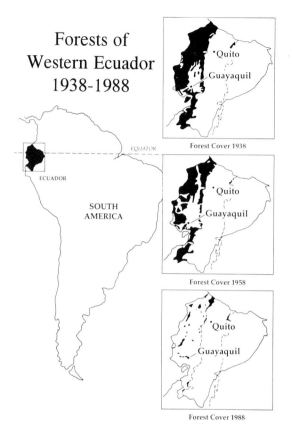

Forests of Western Ecuador 1938-1988

Figure 4.4

In the 35 years between 1950 and 1985, much of the dense forest cover of Central America has been destroyed. (Modified after Leonard, *Natural Resources and Economic Development in Central America*, 1987, p. 118. Reprinted by permission of Transaction Books.)

Forests of Central America, 1950-1985

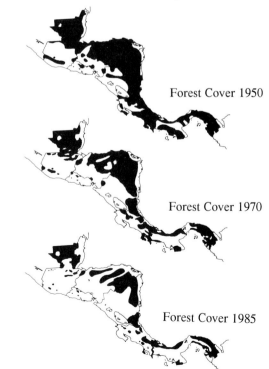

Madagascar has a unique flora and fauna, 80 to 90 percent endemic, one of the highest levels in the world. Twenty-eight of its 30 species of primates (lemurs) are endemic on the island, with four of the five families occurring only there. Eight of the nine species of carnivores, 29 of the 30 species of tenrecs, 237 of the 269 reptiles, and 142 of the 144 species of amphibians are endemic. Three endemic families and 53 species of birds are endemic. There are seven to eight endemic plant families. Palms are abundant with 128 of the 133 native species occurring nowhere else; and over 1,000 species of orchids are native here. Forty-eight percent of the rare plant genera, including 95 percent of the endemic species (species occurring only in Madagascar), grow in the southern spiny forest. Since man arrived on the island only 1,500 to 2,000 years ago, 80 to 90 percent of the forests have been destroyed (Figure 4.5). Extinct animal species include a pygmy hippopotamus, an aardvark, and at least six genera of lemurs. The eastern rainforest is now only one-third or less of its original cover; the upland forests are largely gone and those remaining are greatly modified by exotic trees for timber and fuel. The spiny forest is being fragmented by sisal plantations. The human population is now over 12 million and growth is about 3.1 percent per year (Green and Sussman, 1990; Langrand, 1990; Preston-Mafham, 1991).

The **Western Ghats** of India have 40 percent endemic plants (about 1,500 species) and the forest cover is being removed at a rate of two to three percent a year (Takhtajan, 1986). The wet relict forests of southern **Sri Lanka** have 50 percent endemism and have been

Figure 4.5

Since humans arrived at Madagascar, 80 to 90 percent of the forests of the eastern half of the island have been destroyed. (Modified after Green and Sussman, "Deforestation history of the eastern rain forests of Madagascar from satellite images," in *Science* 248, 1990, p. 213. Reprinted with permission. Copyright © American Association for the Advancement of Science.)

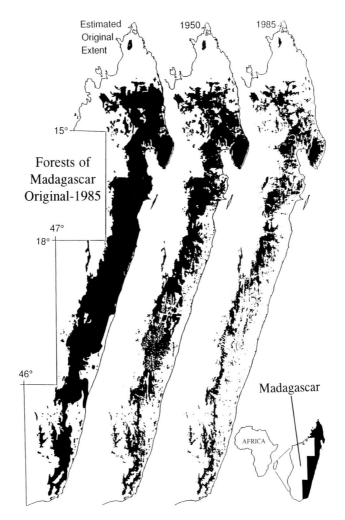

Estimated Original Extent   1950   1985

Forests of Madagascar Original-1985

Madagascar

AFRICA

reduced to less than 10 percent of the original (Takhtajan, 1986). The lower slopes of the **Himalayas** with 39 percent endemism are losing forests through logging and conversion to agriculture. This is an interesting area floristically with a meeting of temperate genera, such as rhododendrons and tropical species, and an abundance of epiphytes (Takhtajan, 1986).

Peninsular **Malaysia** is home to 293 species of mammals, over 1,200 bird, 171 amphibian, 294 reptile, and 15,000 higher plant species, including many epiphytes, and at least one-third of them endemic. Much of this rich forest has been converted to commercial monocultures such as oil palms and rubber (McNeely et al., 1990; Takhtajan, 1986).

Northwestern **Borneo** with 40 percent endemic plants is currently under attack with clear-cutting as well as selective logging. Until recently this had been largely untouched forest. The lower forests here, and also in other hot spots in the Eastern Hemisphere, are dominated by species of the family Dipterocarpaceae. Deforestation in both Malaysia and Borneo is primarily a post–World War II development. The orangutan is no longer in Malaysia, but it persists in Borneo. The disruption of local peoples due to the loss of the forest has been disastrous (Takhtajan, 1986; Wilson, 1992).

The **Philippines** with a rich diversity of plants is essentially a lost cause; planned preserves will occupy only two percent of the land area. Agriculture and logging continue. The monkey-eating eagle, one of the largest birds of prey known, has been reduced to 200 or fewer individuals. The flora combines continental features with oceanic; rhododendrons and pines are native in the cool uplands (Bates, personal communication; Carlquist, 1965, 1974; Takhtajan, 1986; Wilson, 1992).

New Caledonia forests have about 90 percent endemism, containing, as in Madagascar, a unique assemblage of plants, noteworthy for the number of endemic primitive plants. The Araucariaceae are represented by eight species. The forest has been described as looking like a vestige of some strange ancient forest. The harmonic old flora has been disrupted by introductions. The island has no mammals but possesses an apparently flightless bird, the Kagu, with well-developed wings. Disturbances have occurred from mining activities (Carlquist, 1965, 1974; Takhtajan, 1986; Wilson, 1992).

Hawaii, with its several volcanic islands, ranging from the oldest in the northwest to the newest in the southeast, is known for its ecological diversity. Over 90 percent of the flowering plant species are endemic. Extinctions and depletions have been taking place on these islands since the arrival of the Polynesians who introduced pigs, fowl, rats, and dogs, as well as many exotic plants. Introductions by Europeans include sheep, goats, cattle, deer, domestic cats, mongoose, and many exotic birds. The introduction of mosquitoes provided a vector for avian malaria and other foreign diseases to the susceptible local birds. Since World War II the human population has increased significantly, augmented by the increase in tourism. Large areas are being taken over for golf courses, expanding hotel sites particularly along the beaches, and the construction of new highways. Fortunately there is an active program in Hawaii now to preserve the most endangered plants *ex situ* (Carlquist, 1965, 1970, 1974; Cowan, 1976; Scagel, 1975; Takhtajan, 1986; Wagner, Herbst, and Sohmer, 1990).

Three wet forests in South America complete the list of 18 hot spots. The **Colombian Chocó**, extending in the Pacific lowlands and foothills from southern Panama to Ecuador, is one of the richest and least explored floras in the world, with perhaps ten thousand species and 25 percent endemism. In the past 20 years three-fourths of the forest has been cut by timber companies, and as the land is opened by logging roads and clearing, people are moving in, an invasion repeated over and over again, preventing any significant natural regeneration. The Chocó, however, is only a small part of Colombia; the entire country is threatened. It has one of the world's greatest species diversity per unit area. With a land area a little more than one million square kilometers, only 0.77 percent of the Earth's surface, it contains about 10 percent of the Earth's species. Higher plant species number about 50,000, including more orchids (about 3,500 species) than any other country; about 20 percent (over 1,700) of the bird species; and the third highest number of terrestrial vertebrates (2,890 species), including 27 species of neotropical primates (McNeely et al., 1990).

The western **Ecuador** forest is a southward extension of the Chocó with at least 25 percent endemism. Destruction has already been so complete (Figure 4.3) that the few remaining populations of plants and animals may not be able to survive. As long ago as 1931 it was recognized that Ecuador, with wet coastal and Amazonian forests, dry southwestern scrub and woodland, and rich montane and alpine flora, probably had the highest biodiversity of any country in South America, perhaps as many as 30,000 plant species (Dodson and Gentry, 1991; Gentry and Dodson, 1987; Wilson, 1992).

The uplands of **western Amazonia,** extending from Colombia south to Bolivia, are believed to contain the greatest biodiversity of any place on Earth now, rich in endemic species. Gentry (1988) has made several one-hectare inventories of higher plants. In seven plots the range in number of trees was 513 to 858, including 163 to 300 species; lianas (a kind of woody vine) equaling or greater than 10 cm in diameter numbered 14 to 26 in 10 to 17 different species. The number of families represented in each plot ranged from 32 to 58. It has been estimated that as much as 90 percent may be gone by the end of the century, particularly at the higher elevations. The lower flatter areas, dissected by major tributaries, oxbow lakes, and swamps, may survive longer (Gentry and Lopez-Parodi, 1980; Gentry, 1988; Mathias, personal observation; Wilson, 1992).

The Atlantic coastal forest of **Brazil** is isolated from the Amazon forests and has a distinctive flora and fauna, such as the golden lion tamarins. It once covered 100 to 120 million hectares, about 12 percent of the country, extending from Rio Grande do Norte and Ceará in northeastern Brazil in a narrow strip south to Rio Grande do Sul. Only one to five percent of the original forest remains, mostly in small reserves and parks. Urbanization, agriculture, and industrialization are responsible (McNeely et al., 1990; Weinberg, 1992; Wilson, 1992).

HUMAN DIVERSITY

In a discussion of biodiversity we must not forget human societies. Many indigenous peoples and their cultures are disappearing. We have one well-documented case of extinction in Ishi, the last of the Yana tribe of California Indians (Kroeber, 1961). Indigenous peoples who have lived in essentially the same environment for generations have a wealth of knowledge that we are finally beginning to record. They have developed stewardship techniques that we are studying, such as mixed agricultural systems, chinampas well-suited to the wet lowlands of the tropics, and tree gardens, thus saving valuable native trees and maintaining habitats for animals. They have extensive knowledge of the useful plants that is proving helpful to ethnobotanists, who are now systematically recording it. They have about 3,000 languages with only five percent of them spoken by more than a half million people. For example, in South America there are about ten million Quechua speakers. It has been said that "encoded in indigenous languages, customs, and practices may be as much understanding of nature as is stored in the libraries of the modern science" (Durning, 1992). The analysis of the languages can give us information on the history of human migrations. There are still about 5,000 distinct indigenous cultures extant, but they are threatened by violence (such as the recent conflict of the Yanomani in South America with gold miners, and the Penan of Sarawak with timber companies), altered by missionaries, and exploited by entrepreneurs. Their subsistence economics are changed through development, and their homelands are altered by commercial extractors and squatters. These are not just a few people but hundreds of millions. They have been responsible in many ways for the preservation of biodiversity, since they realized its importance to their survival (Balick and Mendelsohn, 1992; Denslow and Paddoch, 1988; Durning, 1992; Holloway, 1993; Nabhan,

1989; Prance and Kallunki, 1984; Redford and Stearman, 1993; Schultes and Raffauf, 1990; Vietmeyer, 1989; Weiskopf, 1993).

■
CONCLUSION

The conservation of biodiversity is a continuing challenge, one that has long been recognized by a few people, pioneers such as George Perkins Marsh, the author of *Man and Nature*. He wrote in 1847 (as quoted by Udall, 1965): "Men now begin to realize what as wandering shepherds they had before dimly suspected, that man has a right to the use, not the abuse, of the products of nature; that consumption should everywhere compensate by increasing production; and that it is false economy to encroach upon a capital, the interest of which is sufficient for our lawful uses." And further:

"The ravages committed by man subvert the relations and destroy the balance which nature had established . . . ; and she avenges herself upon the intruder by letting loose her destructive energies. . . . When the forest is gone, the great reservoir of moisture stored up in its vegetable mould is evaporated. . . . The well-wooded and humid hills are turned to ridges of dry rock . . . and . . . the whole earth, unless rescued by human art from the physical degradation to which it tends, becomes an assemblage of bald mountains, of barren, turfless hills, and of swampy and malarious plains. There are parts of Asia Minor, of Northern Africa, of Greece, and even of Alpine Europe, where the operation of causes set in action by man has brought the face of the earth to a desolation almost as complete as that of the moon. . . . The earth is fast becoming an unfit home for its noblest inhabitant, and another era of equal human crime and human improvidence . . . would reduce it to such a condition of impoverished productiveness, of shattered surface, of climate excess, as to threaten the depravation, barbarism, and perhaps even the extinction of the species."

Unfortunately the advice of Marsh was not heeded until over a hundred years later when a number of agencies and many individuals became actively concerned. We now have worldwide action groups such as the World Wide Fund for Nature and Conservation International, as well as many field scientists (unfortunately not an adequate number) attacking the challenge we face. The United Nations Convention on Biological Diversity (Reid et al., 1993) and the similar statement by a group of forest-dwelling native Amazonians (Redford and Stearman, 1993) are encouraging. Turner (1984) warned us: "When we wipe out the world's rain-forests something will take their place. If we are very clever and very lucky, that something might resemble the farmlands of northern Europe; if we are moderately lucky it might resemble the poor heather moors of the Scottish Highlands; more probably, it will be something like the Sahara." Koshland (1991) has given us our game plan: "Thus, in the long, evolutionary battle, *Homo sapiens* has prevailed, by using its brains, but will win only if it can now use the same brains to limit its victory and ensure its own survival."

■
REFERENCES

Balick, M.J., and Mendelsohn, R. 1992. Assessing the economic value of traditional medicines from tropical rain forests. *Conservation Biology* 6: 128–130.

Barrett, S.C.H., and Kohn, J.R. 1991. Genetic and evolutionary consequences of small populations size in plants: Implications for conservation. In: D.O. Falk and K.E. Holzinger (eds.), *Genetics and Conservation of Rare Plants* (Oxford, UK: Oxford University Press), pp. 3–30.

Carlquist, S. 1965. *Island Life* (Garden City, NJ: Natural History Press).

———. 1970. *Hawaii, a Natural History* (New York: Natural History Press).

———. 1974. *Island Biology* (New York: Columbia University Press).

Cowan, I.M. 1976. Biota Pacifica 2000. In: R.F. Scagel (ed.), *Mankind's Future in the Pacific* (Vancouver: University of British Columbia Press), pp. 86–98.

Del Moral, R., and Wood, D.M. 1993. Early primary succession on a barren volcanic plain at Mount St. Helens, Washington. *American Journal of Botany* 80: 981–991.

Denslow, J.S., and Paddoch, C. 1988. *People of the Tropical Rain Forest* (Berkeley, CA: University of California Press).

Dodson, C.H., and Gentry, A.H. 1991. Biological extinction in western Ecuador. *Annals of the Missouri Botanical Garden* 78: 273–295.

Durning, A.B. 1992. *Guardians of the Land: Indigenous Peoples and the Health of the Earth.* Worldwatch Paper 112 (Washington, DC: Worldwatch Institute).

Ehrlich, P.R., and Murphy, D.D. 1987. Conservation lessons from long-term studies of checkerspot butterflies. *Conservation Biology* 1: 122–131.

Erwin, T.L. 1991. How many species are there?: Revisited. *Conservation Biology* 5: 330–333.

Falk, D.O., and Holzinger, K.E. (eds.). 1991. *Genetics and Conservation of Rare Plants* (Oxford, UK: Oxford University Press).

Frayer, W.E., Peters, D.D., and Pywell, H.R. 1989. *Wetlands of the California Central Valley, Status and Trends, 1939 to mid 1980's* (Portland, OR: U.S. Fish and Wildlife Service Region).

Gentry, A.H. 1988. Tree species richness of upper Amazonian forests. *Proceedings of the National Academy of Sciences USA* 85: 156–159.

Gentry, A.H., and Dodson, C.H. 1987. Diversity and biogeography of neotropical vascular epiphytes. *Annals of the Missouri Botanical Garden* 74: 205–233.

Gentry, A.H., and Lopez-Parodi, J. 1980. Deforestation and increased flooding of the upper Amazon. *Science* 210: 1354–1356.

Green, G.M., and Sussman, R.W. 1990. Deforestation history of the eastern rain forests of Madagascar from satellite images. *Science* 248: 212–215.

Groombridge, B. (ed.). 1992. *Global Biodiversity, A Report Compiled by the World Conservation Monitoring Centre* (London: Chapman & Hall).

Holloway, M. 1993. Sustaining the Amazon. *Scientific American* 269 (1): 90–91.

Huenneke, L. 1991. Ecological implications and genetic variation in plant populations. In: D.O. Falk and K.E. Holzinger (eds.), *Genetics and Conservation of Rare Plants* (Oxford, UK: Oxford University Press), pp. 31–61.

Janzen, D.H. 1986. Tropical dry forests: The most endangered major tropical ecosystem. In: E.O. Wilson (ed.), *Biodiversity* (Washington, DC: National Academy Press).

———. 1988. Management of habitat fragments in a tropical dry forest: Growth. *Annals of the Missouri Botanical Garden* 75: 105–116.

Koshland, D.E., Jr. 1991. Preserving biodiversity. *Science* 253: 717.

Kroeber, T. 1961. *Ishi.* (Berkeley, CA: University of California Press).

Langrand, O. 1990. *Birds of Madagascar* (New Haven, CT: Yale University Press).

Larson, D. 1993. The recovery of Spirit Lake. *American Scientist* 8: 166–177.

Ledig, F.T. 1986. Heterozygosity, heterosis, and fitness in outbreeding plants. In: M.E. Soule (ed.), *Conservation Biology: The Science of Scarcity and Diversity* (Sunderland, MA: Sinauer Associates), pp. 77–104.

Leonard, H.J. 1987. *Natural Resources and Economic Development in Central America* (New Brunswick, NJ: Transaction Books).

Lipman, P.W., and Mullineaux, D.R. (eds.). 1981. *The 1980 Eruption of Mount St. Helens, Washington.* Geological Survey Professional Paper 1250 (Washington, DC: U.S. Government Printing Office).

Lovejoy, T. 1992. Infinite variety—a rich diversity of life. In: L. Silcock (ed.), *The Rainforests* (San Francisco: Chronicle Books), pp. 35–38.

McNeely, J.A., Miller, K.R., Reid, W.V., et al. 1990. *Conserving the World's Biological Diversity* (Gland, Switzerland: International Union for the Conservation of Nature and Natural Resources).

Mathias, M.E., and Moyle, P. 1992. Wetlands and aquatic habitats. In: G. Gall (ed.), *Agriculture, Ecosystems and Environment* (Amsterdam: Elsevier), pp. 165–176.

Matthiessen, P. 1959. *Wildlife in America* (New York: Viking Press).

May, R.M. 1988. How many species are there on earth? *Science* 241: 1441–1449.

———. 1992. How many species inhabit the earth? *Scientific American* 267 (4): 42–48.

Myers, N. 1988. Threatened biotas: "Hotspots" in tropical forests. *Environmentalist* 8 (3): 1–20.

———. 1990. The biodiversity challenge: Expanded hot spots analysis. *Environmentalist* 10 (4): 243–256.

Nabhan, G.P. 1989. *Enduring Seeds: Native American Agriculture and Wild Plant Conservation* (San Francisco: North Point Press).

Norton, D.A. 1991. *Trilepidea adamsii:* An obituary for a species. *Conservation Biology* 5: 52–57.

Ogden, J.C. 1992. The impact of Hurricane Andrew on the ecosystem of south Florida. *Conservation Biology* 6: 488–490.

Prance, G.T., and Kallunki, J.A. (eds.). 1984. *Ethnobotany in the Neotropics* (Bronx, NY: New York Botanical Garden).

Preston-Mafham, K. 1991. *Madagascar: A Natural History* (Oxford, UK: Facts on File).

Redford, K.H., and Stearman, A.M. 1993. Forest-dwelling native Amazonians and the conservation of biodiversity: Interests in common or in collision? *Conservation Biology* 7: 248–255.

Reid, W.V., Laird, S.A., Meyer, C.A., et al. 1993. *Biodiversity Prospecting: Using Genetic Resources for Sustainable Development* (Washington, DC: World Resources Institute).

Robinson, J.G., and Redford, K.H. (eds.). 1991. *Neotropical Wildlife Use and Conservation* (Chicago, IL: University of Chicago Press).

Scagel, R.F. 1975. *Mankind's Future in the Pacific* (Vancouver: University of British Columbia Press).

Schultes, R.E., and Raffauf, R.F. 1990. *The Healing Forest: Medicinal and Toxic Plants of the Northwest Amazonia* (Portland, OR: Dioscorides Press).

Soulé, M.E., and Kohn, K. A. 1989. *Research Priorities for Conservation Biology* (Washington, DC: Island Press).

Takhtajan, A. 1986. *Floristic Regions of the World* (Berkeley, CA: University of California Press).

Terborgh, J. 1992. Why American songbirds are vanishing. *Scientific American* 266 (5): 98–104.

Turner, J.R.G. 1984. Extinction as a creative force: The butterflies of the rain-forest. In: A.C. Chadwick and S.L. Sutton (eds.), *Tropical Rain-Forest* (Leeds, UK: Leeds Philosophical and Literary Society), pp. 195–204.

Udall, S.L. 1965. *The Quiet Crisis* (New York: Holt, Rinehart and Winston).

Varley, J.A. 1979. Physical and chemical soil factors affecting the growth and cultivation of endemic plants. In: H. Synge and H. Townsend (eds.), *Survival or Extinction* (Kew, UK: The Bentham-Moxon-Trust, Royal Botanic Gardens), pp. 197–205.

Vietmeyer, N. D. (ed.). 1989. *Lost Crops of the Incas* (Washington, DC: National Academy Press).

Wagner, W.L., Herbst, D.R., and Sohmer, S.H. 1990. *Flora of Hawaii* (Honolulu: University of Hawaii Press).

Weinberg, S. 1992. Return of a native. *Pacific Discovery* 45 (2): 8–14.

Weiskopf, J. 1993. Healing secrets in a shaman's forest. *Americas* 45 (4): 42–47.

Wilson, E.O. 1992. *The Diversity of Life* (Cambridge, MA: Harvard University Press).

CHAPTER 5

GENDER BIAS AND THE SEARCH FOR A SUSTAINABLE FUTURE

■

Jodi L. Jacobson*

■

INTRODUCTION

The women of Sikandernagar, a village in the Indian state of Andhra Pradesh, work three shifts per day. Waking at 4:00 A.M., they light fires, milk buffaloes, sweep floors, fetch water, and feed their families. From 8:00 A.M. until 5:00 P.M., they weed crops for a meager wage. In the early evening they forage for branches, twigs, and leaves to fuel their cooking fires, for wild vegetables to nourish their children, and for grass to feed the buffaloes. Finally, they return home to cook dinner and do evening chores. These women spend twice as many hours per week working to support their families as do the men in their village. However, they do not own the land on which they labor, and every year, for all their effort, they find themselves poorer and less able to provide what their families need to survive (Mies, 1986).

As the 20th century draws to a close, some three billion people—more than half the Earth's population—live in the subsistence economies of the Third World (Durning, 1989; UNDP, 1991). The majority of them find themselves trapped in the same downward spiral as the women of Sikandernagar.

In the not-so-distant past, **subsistence farmers** and forest dwellers were models of ecologically sustainable living, balancing available resources against their numbers. Today, however, the access of subsistence producers to the resources on which they depend for survival is eroding rapidly. As their circumstances grow more and more tenuous, pressures on the forests and croplands that remain within their grasp grow increasingly acute. Yet in an era when sustainable development has become a global rallying cry, most governments and international development agencies seem oblivious to this dilemma.

The reason is brutally simple: Women perform the lion's share of work in subsistence economies, toiling longer hours and contributing more to family income than men do. Yet

*Health and Development Policy Project, 1730 Rhode Island Avenue, NW, Suite 712, Washington, DC 20036. Reprinted from *State of the World*, Worldwatch Institute 1993. (New York: W. W. Norton). An expanded version of this chapter appears in Jacobson, J.L. 1992. *Gender Bias: Roadblock to Sustainable Development.* Worldwatch Paper 110 (Washington, DC: Worldwatch Institute).

in a world where economic value is computed in monetary terms alone, women's work is not counted as economically productive when no money changes hands.

Women are viewed as "unproductive" by government statisticians, economists, development experts, and even their husbands. A huge proportion of the world's real productivity therefore remains undervalued, and women's essential contributions to the welfare of families and nations remain unrecognized. So while the growing scarcity of resources within subsistence economies increases the burden on women and erodes their productivity, little is being done to reverse the cycle.

Ironically, by failing to address the pervasive gender bias that discounts the contributions of women, development policies and programs intended to alleviate impoverishment—and the environmental degradation that usually follows—actually are making the problem worse.

Gender bias is a worldwide phenomenon, afflicting every social institution from individual families to international development organizations. But it is especially pernicious in the Third World, where most of women's activity takes place in the nonwage economy for the purpose of household consumption. In Sikandernagar, for example, women earn less than half the amount men do for the same work. Because their cash income is not enough to buy adequate supplies of food and other necessities (which they are responsible for obtaining one way or another), they work additional hours to produce these goods from the surrounding countryside (Mies, 1986).

In most societies, gender bias compounds—or is compounded by—discrimination based on class, caste, or race. It is especially pervasive in the poorest areas of Africa, Asia, and Latin America, where it ranges from the exclusion of women from development programs to wage discrimination and systemic violence against females. In its most generic form, this prejudice boils down to grossly unequal allocation of resources—whether of food, credit, education, jobs, information, or training.

Gender bias is thus a primary cause of poverty, because in its various forms it prevents hundreds of millions of women from obtaining the education, training, health services, child care, and legal status needed to escape from poverty. It prevents women from transforming their increasingly unstable subsistence economy into one not forced to cannibalize its own declining assets. And it is also the single most important cause of rapid population growth. Where women have little access to productive resources and little control over family income, they depend on children for social status and economic security. The greater competition for fewer resources among growing numbers of poor people accelerates environmental degradation. Increased pressure on women's time and labor in turn raises the value of children—as a ready labor force and hedge against an uncertain future. The ensuing high rates of population growth become part of a vicious cycle of more people, fewer resources, and increasing poverty. A necessary step in reducing births voluntarily, then, is to increase women's productivity and their control over resources.

■

THE DIMENSIONS OF GENDER BIAS

Implicit in the theory and practice of conventional economic development are three assumptions that are influenced by sex differences—and that reinforce the biases. One assumption is that within a society, both men and women will benefit equally from economic growth. The second is that raising men's income will improve the welfare of the whole family. The third is that within households, the burdens and benefits of poverty and wealth will be distributed equally regardless of sex. Unfortunately, none of these assumptions holds true.

The first assumption—that economic growth is gender blind—is rarely challenged. However, as economies develop, existing gender gaps in the distribution of wealth and in access to resources usually persist, and in many cases grow worse. From the 1950s through the early 1980s, for example, worldwide standards of living as measured by widely used basic indicators—including life expectancy, per capita income, and primary school enrollment—rose dramatically. Yet women never achieved parity with men, even in industrial countries.

According to the **Human Development Index** prepared by the United Nations Development Programme, which gauges the access people have to the resources needed to attain a decent standard of living, women lagged behind men in every country for which data were available. The differences were least pronounced in Sweden, Finland, and France, where measures of women's level of access as a share of men's passed 90 percent. They were most pronounced in Swaziland, South Korea, and Kenya, where women had less than 70 percent the access that men did (UNDP, 1991).

Not only do women not automatically benefit from economic growth; they may even fall further behind. Unless specific steps are taken to redress inequity, gender gaps often increase over time—especially where access to resources is already highly skewed. This has happened, for example, with literacy. In 1985, 60 percent of the adult population worldwide was able to read, compared with about 46 percent in 1970—clearly a significant improvement. Literacy rose faster among men than among women, however, so the existing gender gap actually widened. Between 1970 and 1985, the number of women unable to read rose by 54 million (to 597 million), while that of men increased by only four million (to 352 million). These numbers reflect females' much lower access to education in developing countries (UNDIESA, 1991).

The second assumption—that social strategies to raise men's income by increasing their access to productive resources will lead directly to improvements in total family welfare— is also not supported by the evidence. It may seem reasonable to assume that each dollar of income earned by a poor man in Bangladesh, Bolivia, or Botswana would go toward bettering the lot of his wife and children. Indeed, development programs have been built on the premise that what is good for men is good for the family. But in many areas this is patently not the case, because it is women who effectively meet the largest share of the family's basic needs, and because men often use their income to purchase alcohol, tobacco, or other consumer products.

Generally speaking, men in subsistence economies have fewer responsibilities than women to produce food and other goods solely for household consumption. While a woman labors to produce food for her children and family, her husband may focus his energies on developing a business or pursuing interests that do not include his wife and children.

In much of sub-Saharan Africa, for instance, both men and women plant crops, but they do so with different goals. Husbands and wives maintain separate managerial and financial control over the production, storage, and sale of their crops. Men grow cash crops and keep the income from them—even though their wives still do the weeding and hoeing. Women, by contrast, use their land primarily for subsistence crops to feed their families. They are also expected to provide shelter, clothing, school fees, and medical care for themselves and their children, and so must earn income to cover what they cannot produce or collect from the village commons land. Given adequate acreage, high yields, or both, women do plant and market surplus crops to earn cash. When land is scarce or the soil poor, they sell their labor or put more time into other income-producing activities (Cleaver and Schreiber, 1992; Davidson, 1988; ILO, 1984).

Because responsibilities for securing the goods needed for household consumption often fall to the woman, even an increase in the income of a male within a household may not mean an increase in total consumption by family members. As subsistence economies become increasingly commercialized, for example, men whose families are below the poverty line often spend any additional cash income to raise the productivity of their own crops, and sometimes to increase their personal consumption. In Africa, according to one World Bank report, "it is not uncommon for children's nutrition to deteriorate while wrist watches, radios, and bicycles are acquired by the adult male household members." The connection between malnutrition and the diversion of income by males to personal consumption has also been found in Belize, Guatemala, Mexico, and throughout the Indian subcontinent (Cleaver and Schreiber, 1992; Carr, 1985; Blumberg, 1990; Acsadi and Johnson-Acsadi, 1987).

In fact, contrary to conventional assumptions, women are the main breadwinners in a large share of families throughout the Third World. They contribute proportionately more of their cash income to family welfare than men do, holding back less for personal consumption. A study in Mexico found that wives accounted for 40 percent or more of the total household income, although their wage rates were far lower than their husbands'. The women contributed 100 percent of their earnings to the family budget, while husbands contributed at most 75 percent of theirs. Similar discrepancies in the amount of money contributed have been found to be virtually universal throughout the developing world (Blumberg, 1990).

Moreover, studies in every region of the Third World confirm that it is the mother's rather than the father's income or food production—and the degree of control she maintains over that income—that determines the relative nutrition of children. In Guatemala, for example, the children of women earning independent incomes had better diets than those of women who were not earning their own money or who had little control over how their husbands' earnings were spent. Women who retain control over income and expenditures spend more not only on food but also on health care, school expenses, and clothing for their children. Similar patterns have been found in studies from the Dominican Republic, Ghana, India, Kenya, Peru, and the Philippines (Blumberg, 1990; Agarwal et al., 1990; Acsadi and Johnson-Acsadi, 1987).

Differences in the responsibilities and workloads of men and women within subsistence economies can also affect family welfare. A project in the Indian state of West Bengal, for example, gave villagers conditional access to trees on private land. The "lops and tops" of trees were to be reserved for women's needs, while men were to harvest the timber for cash on a sustainable basis. In response to offers from a contractor, however, the men sold the trees for a lump sum. Women obtained little fuel (Molnar and Schreiber, 1989).

The third assumption—that within poor households resources will be distributed equally regardless of sex—may seem so obvious as to be beyond question. But even when a man's income is used to improve his family's, it may improve the welfare of males at the expense of females. In many cultures, a family's resources are distributed according to the status of household members, rather than according to their needs. Men and boys fare far better than women and girls. In India, for instance, studies show that in many states sons consistently receive more and better food and health care than their sisters. Consequently, far more girls than boys die in the critical period between infancy and age five. And with the exception of girls aged 10 to 14, Indian females die from preventable causes at far higher rates than males do through age 35 (Chatterjee, 1991; Ghosh, 1991).

Basic indicators of caloric intake and life expectancy measured by the Indian government's 1991 census reveal a growing gender gap in several states since 1980. In fact, contrary to sex ratios found in most countries, the ratio of women to men in India has actually been declining since the early part of the century. There are now only 929 women for every 1,000 men, compared with 972 in 1901. Dr. Veena Mazumdar, director of the Delhi-based Centre for Women's Development Studies, notes that "the declining sex ratio is the final indicator that registers [that] women are losing out on all fronts—on the job market, in health and nutrition and economic prosperity" (Chatterjee, 1991; Ram, 1991; Census Commissioner, 1991).

Evidence of similar patterns of discrimination in the allocation of household resources has been found in Bangladesh, Nepal, Pakistan, throughout the Middle East and North Africa, and in parts of sub-Saharan Africa. Harvard economist and philosopher Amartya

Sen calculates that 100 million women in the developing world are "missing," having died prematurely from the consequences of such gender bias (Acsadi and Johnson-Acsadi, 1987; Jacobson, 1991; Sen, 1990).

Because of these patterns, argues Bina Agarwal, professor of agricultural economics at the Institute of Economic Growth in Delhi, "existing poverty estimates need revision." The current practice is to first identify poor households by specified criteria and then calculate the total numbers, the assumption being that all members are equally poor. However, Agarwal argues, this reveals little about the relative poverty of men and women. The differences in the distribution of resources within households mean there are poor women in households with cash incomes or consumption levels above the poverty line. Conversely, there are nonpoor men in households below the poverty line (Agarwal, 1988).

Globally, much of this discrimination against females in families and societies stems from another form of gender gap—the huge disparity between the real economic and social benefits of women's work and the social perception of women as unproductive.

In every society, women provide critical economic support to their families, alone or in conjunction with spouses and partners, by earning income—in case or in kind—in agriculture, in formal and informal labor markets, and in emerging international industries, such as the manufacture of semiconductors. United Nations data indicate that, on average, women work longer hours than men in every country except Australia, Canada, and the United States. Hours worked earning wages or producing subsistence goods are rarely offset by a reduction of duties at home. Time allocation studies confirm that women throughout the world maintain almost exclusive responsibility for child care and housework. Moreover, disparities in total hours worked are greatest among the poor: In developing countries, women work an average of 12 to 18 hours a day—producing food, managing and harvesting resources, and working at a variety of paid and unpaid activities—compared with 8 to 12 hours on average for men (UNDIESA, 1991; Agarwal et al., 1990).

In subsistence economies, measuring work in terms of the value of goods produced and time spent shows that women usually contribute as much as or more than men to family welfare. The number of female-headed households is growing. However, "even where there is a male earner," notes World Bank consultant Lynn Bennett, "women's earnings form a major part of the income of poor households" (Bennett, 1989).

The low valuation of women's work begins with the fact that in developing countries, most of women's activity takes place in the **nonwage economy** for the purpose of household consumption—producing food crops, collecting firewood, gathering fodder, and so on. "Income generation" of this type is critically important; indeed, the poorer the family, the more vital is the contribution of women and girls to the essential goods that families are unable to buy with cash. But in the increasingly market-oriented economies of the Third World, work that does not produce cash directly is heavily discounted (Bennett, 1989; Agarwal et al., 1990; Acsadi and Johnson-Acsadi, 1987).

Low valuation is further reinforced by women's institutionally enforced lack of control over physical resources. In most subsistence economies, females have few legal rights regarding land tenure, marital relations, income, or social security. In a world where control over land confers power, the value of wives' and mothers' contributions in subsistence economies also is discounted because these are directed mainly at day-to-day sustenance and do not yield such visible assets. The "invisible" nature of women's contributions feeds into the social perception that they are "dependents" rather than "producers." Indeed, the tendency at every level of society seems to be to play down the importance of female contributions to family income, which anthropologist Joke Schrijvers, cofounder of the Research and Documentation Centre on Women and Autonomy in the Netherlands, attributes to the "ideology of the male breadwinner" (Schrijvers, 1988).

The ideology appears to be universal. And rather than combating the idea that women's work has low economic value, governments and international development agencies have tacitly condoned it. Thus, despite overwhelming evidence to the contrary, these institutions persist in counting women as part of the dependent or "nonproductive" portion of the population.

This bias is then perpetuated by government record keeping practices: Official definitions of what constitutes "work" often fail to capture a large share of women's labor. In India, conventional measures based on wage labor showed that only 34 percent of Indian females are in the labor force, as opposed to 63 percent of males. But a survey of work patterns by occupational categories including household production and domestic work revealed that 75 percent of females over age five are working, compared with 64 percent of males. In a study of Nepalese villages, estimates of household income based only on wages earned put the value of female contributions at 20 percent. Taking account of subsistence production, however, brought this contribution to 53 percent. And in a study of women in the Philippines, "full income" contributions were found to be twice as high as marketed income (Bennett, 1989; Chatterjee, 1991; UNDIESA, 1991).

Given such distorted pictures of their national economies, it is not surprising that policy makers in virtually every country invest far less in female workers than in males. Moreover, international development assistance agencies, staffed mostly by men with a decidedly Western view of the world, have based their decisions on the erroneous premise that what is good for men is good for the family. And because most strategists neither integrate women into their schemes nor create projects that truly address women's economic needs, development efforts aimed at raising productivity and income often bypass women altogether.

Ignoring the full value of women's economic contributions cripples efforts to achieve broad development goals. Lack of investment results in lower female productivity. Coupled with persistent occupational and wage discrimination, this prevents women from achieving parity with men in terms of jobs and income, and it leads to further devaluation of their work. The omnipresence of this bias is a sign that virtually every country is operating far below its real economic potential.

Current measures of economic development tell little about how the benefits of that development will be distributed. Higher aggregate levels of agricultural production, for example, do not necessarily imply lower levels of malnutrition. A rising gross national product does not always produce a decline in the incidence of poverty or an improvement in equity. And a real increase in the health budget of a country does not automatically lead to better access to primary health care among those most in need of it. With any project or investment, it is important to ask: Whose income is rising? Whose opportunities are increasing?

SUSTENANCE FROM THE COMMONS

Because women in rural subsistence economies are the main providers of food, fuel, and water and the primary caretakers of their families, they depend heavily on community-owned croplands, grasslands, and forests to meet their families' needs. The widespread depletion and degradation of these resources have led to equally widespread impoverishment of subsistence families throughout Africa, Asia, and Latin America.

In rural areas, both men and women engage in agriculture, but women are the major producers of food for household consumption. In sub-Saharan Africa, women grow 80 percent of the food destined for their households. Women's labor produced 70 to 80 percent of food crops grown on the Indian subcontinent and 50 percent of the food domestically consumed in Latin America and the Caribbean. In all regions, roughly half of all cash crops are cultivated by women farmers and agricultural laborers (Jacobson, 1988; Gittinger, 1990; Russo et al., 1989; Chatterjee, 1991).

By custom, labor contributions are divided by sex. In sub-Saharan Africa, for example, males generally clear and till the land, while females are expected to do the bulk of the hoeing, weeding, and harvesting of crops, the processing of food, and various other subsistence activities. Women, therefore, perform the majority of the work in African agriculture. Similar patterns in the division of labor are found throughout the subsistence economies of Asia and Latin America (ILO, 1984; Deere and Leon de Leal, 1982; Poats et al., 1988).

Using traditional methods, women farmers have been quite effective in **conserving soil**. Given access to appropriate resources, they employ managed fallowing (allowing land to rest between plantings), crop rotation, intercropping, mulching, and a variety of other soil conservation and enrichment techniques. They have also played a leading role in maintaining crop diversity. In sub-Saharan Africa, for instance, women cultivate as many as 120 different plants in the spaces alongside men's cash crops. In the Andean regions of Bolivia, Colombia, and Peru, women develop and maintain the seed banks on which food production depends (Abramovitz and Nichols, 1992).

Faced with the endemic insecurity of their situation, women have evolved techniques to make efficient use of all available resources—planting a diverse array of crops, collecting wild fruits and vegetables, maintaining farm animals, and earning whatever cash income they can to ensure a measure of food security. Although cultivating food is obviously difficult or impossible for women in families with little or no land, studies show the land-poor to be highly resourceful in devising ways to meet their families' needs. Solutions used include drawing more heavily on products gathered from commons, expanding their workloads, and hiring themselves out as laborers in exchange for grain or cash (Gittinger, 1990; Rocheleau, 1988; Williams, 1991a).

Women in subsistence economies also are active managers of **forest resources** and traditionally play the leading role in their conservation. Forests provide a multitude of products to households. They are, for example, a major source of fuel, without which none of the food grown and harvested could be cooked, or many other essential tasks be carried out. In fact, lack of fuel to cook available food is itself a cause of malnutrition in some areas. "It's not what's in the pot, but what's under it, that worries you," say women in the fuel-deficit areas of India (Rocheleau, 1988; Williams, 1991a; Molnar and Schreiber, 1989; Agarwal, 1988).

The dependence of subsistence households on biomass—including wood, leaves, and crop residues—as the traditional form of domestic energy remains widespread. Seventy-five percent of all household energy in Africa is derived from biomass, for example. Women also use biomass fuel to support innumerable private enterprises, such as food processing and pottery, from which they gain cash income (Williams, 1991b).

Women depend heavily on the availability of **nonwood forest products**, too. They collect plant fibers, medicinal plants and herbs, seeds used in condiments, oils, resins, and a host of other materials used to produce goods or income for their families. The fruits, vegetables, and nuts widely gathered as supplements to food crops are important sources of protein, fats, vitamins, and minerals not found in some staple crops. In times of drought, flood, or famine, these gathered foods have often made the difference between life and death (Agarwal, 1988; Molnar and Schreiber, 1989; Williams, 1991b).

In most subsistence economies, forest products are key sources of jobs and income. Throughout the Third World, women make up a large share of the labor force in forest industries, from nurseries to plantations, and from logging to wood processing. In hard times, landless and underemployed female agricultural laborers often fall back on the collection of nontimber forest products to generate cash (Molnar and Schreiber, 1989).

These nonwood forest products make substantial contributions to local and national economies. Although unrecognized or unrecorded by national statistics, these activities often contribute more to national income than wood-based industries do. A report by World Bank researchers Augusta Molnar and Gotz Schreiber estimates that in India, for example, nontimber products account for two-fifths of domestic forest revenues and three-fourths of net export earnings from forestry products (Molnar and Schreiber, 1989).

In the true spirit of sustainability, female subsistence producers appear to be as careful in conserving forests as they are reliant on using them. With traditional methods of extraction, women in Africa and Asia obtain their fuel from branches and dead wood (often supplemented with crop residues, dried weeds, or leaves) rather than live trees. Seventy-five percent of domestic fuel collected by women in northern India is in this form. Women are also the chief repositories of knowledge about the use and management of trees and other forest products (Williams, 1991b).

In surveys, women have consistently cited the ecosystem services provided by forests, such as their critical role in replenishing freshwater supplies, as reasons for their preservation. In fact, research on communal resource management systems—the "commons" on which women depend so heavily in subsistence economies—shows them to be more effective at protecting and regenerating the environment than management approaches taken by either the state or private landowners (Agarwal, 1991; Agarwal and Narain, 1990).

The reasons are obvious: Commons are as indispensable to land-poor women in subsistence economies as these women are to the maintenance of the commons. A study of 12 semiarid districts of India in the early 1980s, for instance, showed that 66 to 84 percent of total domestic fuel needs of both the land-poor (those with less than two hectares of dry land equivalent) and the landless were derived from these commons areas, as opposed to 8 to 32 percent of the nonpoor. The poorest households also relied heavily on these lands for grazing (see Table 5.1). Even in the northwest of the country, where **Green Revolution** technologies have been widely applied, commons land accounted for the bulk of foods used to supplement the cereals bought by the poor or earned in kind (Agarwal, 1991).

Commons lands constitute the one resource, apart from their children, that women traditionally have had access to relatively unfettered by the control of men. Unfortunately, women's access to these lands and the goods they yield is fast diminishing. The results are already evident in declining food security among subsistence households.

CASH CROPS VERSUS FOOD SECURITY

Three interrelated trends, all set in motion or perpetuated by the agricultural strategies of low-income countries since the 1950s, have particularly damaged the ability of rural women to produce or procure enough food and in fact are a product of the increasing emphasis on cash crops.

Table 5.1

Share of total income reaped from village commons by poor and nonpoor families in seven states of India, 1985. (Data from Jodha, 1990.)

Commodity	Share of Total Income (Percent)	
	Poor Families	**Nonpoor Families**
Firewood	91–100	–
Domestic Fuel Supplies	66–84	8–32
Grazing Needs	70–90	11–42
Gathered Food	10	–

First, **large amounts of land once jointly owned and controlled by villagers—and accessible to women—have shifted into the hands of government agencies and private landowners.** Second, the distribution of resources on which cash crop agriculture depends heavily—including land, fertilizers, pesticides, irrigation, and hybrid seeds—has reflected persistent gender bias. Third, the mechanization of agriculture has reduced or replaced the labor traditionally done by men, but increased that done by women without raising their income.

The first of these trends has been hastened by development strategies that, as noted earlier, make false assumptions about who benefits from gross economic gains. Thus, while shifting ownership to government agencies and private landowners may improve agricultural output in the aggregate, it fails to address critical differences between men's and women's farming responsibilities.

Despite their major role in food production, women farmers rarely have land of their own. In the patrilineal cultures found in Bangladesh, India, Pakistan, Latin America, and much of sub-Saharan Africa, women gain access to land only through their husbands or sons. In the past at least, customary laws have afforded women some security of land tenure (Agarwal, 1991; Davidson, 1988; Poats et al., 1988).

In the "common property" systems of precolonial Africa and Asia, access to the resources needed by men and women to fulfill their respective family obligations was determined by sex. Consequently, while women could rarely "own" land, as members of a community they usually had equal rights to use it in accordance with their family's needs. Under traditional systems operating in parts of southern Ghana, for instance, women had rights to land as members of a lineage; they applied to the male head of their lineage for the acreage needed for food production, which was allocated according to a family's size and needs (Cleaver and Schreiber, 1992).

These customary laws and practices have been deteriorating since the beginning of the colonial period. The process accelerated with the independence of African countries as the result of several influences. Declining mortality rates led to rapid population growth and increased demand for cropland. Increasing migration, and governments' policies encouraging the acquisition of land by the state and individual producers, put previously accessible land beyond the reach of impoverished women (Cleaver and Schreiber, 1992).

Common property resources traditionally controlled by a community have either been privatized or turned into open access systems over which virtually no one has control. In most regions, privatization of land was an explicit policy: Investments by governments, donor agencies, and multinational corporations directly encouraged the shift of land away from subsistence to cash crops.

In a 1992 study, *The Population, Agriculture and Environment Nexus in Sub-Saharan Africa,* World Bank researchers Kevin Cleaver and Gotz Schreiber state that African governments allowed customary law to guide use of some of the land, "while arbitrarily allocating other land to private investors, political elites, and public projects." In Latin America, there is little commons land left; most of it long ago was privatized and shifted into cash crops. The highly skewed land ownership patterns now found in countries like Brazil, where two percent of the farmers own 57 percent of the arable land and more than half of all agricultural families own none, are but the most obvious legacies of such shifts (Prosterman and Riedinger, 1987).

As a result of privatization favoring male landholders, the amount and quality of land available to women food producers in the Third World is declining. Legal and cultural obstacles prevent women from obtaining title to land and, therefore, from participating in cash crop schemes. Land titles invariably are given to men because governments and international agencies routinely identify them as heads of their households, regardless of whether or not they actually support their families. Women's rights to land are now subject to the wishes of their husbands or the whims of male-dominated courts and community councils.

In Zambia, for example, women continue to be discriminated against in the allocation of land despite the 1975 Land Act guaranteeing women equal access. Researcher Mabel Milimo of the University of Zambia points out that women's access remains limited by the control men have over distributing land. The act vests all land in the president, who in turn delegates his powers of allocation to district councils and other local bodies. Although women farmers outnumber male ones, the councils are made up of men and often require a husband's consent for a married woman to receive land. Milimo points out that "in most cases, the husbands are reluctant . . . because they prefer the wife to work on their . . . cash crops" (Milimo, 1987).

Dianne Rocheleau, a geographer at Clark University in Massachusetts, notes that the erosion of women's customary rights under modern legal reforms is widespread throughout Africa, Asia, and Latin America. In Southeast Asia and the Indian subcontinent, for instance, at least 70 percent of the female labor force is engaged in food production—yet fewer than 10 percent of women farmers in India, Nepal, and Thailand now own land (Agarwal, 1988). Rocheleau contends that both modern and traditional laws tend to be interpreted in favor of male ownership and control, and that even where reforms have provided for equality in most spheres, "strong gender inequality often exists with respect to [allocating] agricultural land." Either way, a woman's access to land is tied to her marital status and the number of sons she bears (Rocheleau, 1988).

The second trend—**discrimination in the allocation of agricultural resources**—follows directly from the first. Agricultural development schemes promoted by governments and development agencies encourage expansion of cash crop operations by offering market incentives, improved agricultural technologies, credit, seeds, and the like—with access to these resources dependent mainly on the use of land as collateral. Without ownership and control over land, women are further disadvantaged because they cannot compete in cash crop schemes and have fewer resources with which to produce food crops (Cleaver and Schreiber, 1992; Davidson, 1988).

The focus of agricultural research and development, too, has favored cash crops over food ones to the detriment of women and of family nutrition. This has not only diminished the relative productivity of food crops, but it has also increased the perception that cash crops are more "valuable." The spread of high-yielding seeds has nearly doubled maize yields in Africa since 1950, for example, but cultivating hybrid maize, a cash crop, is expensive. New seeds must be bought each year, and the crop demands repeated applications of fertilizer. Subsistence farmers who lack credit can afford neither the seeds nor the fertilizer. On the other hand, the development of high-yielding varieties of millet and sorghum has lagged far behind, even though these traditional food crops are considerably more drought-resistant and nutritionally balanced than maize (Dey, 1984).

Vegetables, beans, cassava, and fruit crops, all important to family nutrition in subsistence economies, also have been neglected. These are usually not considered "important" because they are not perceived as being related to food trade. International Labour Organization consultant Ingrid Palmer argues that such policies are shortsighted because growing more of these crops would not only diminish the need for imports, it would provide a chief source of cash to women who sell them in local markets. Furthermore, because many of these "minor" crops keep well, they are available to families for a longer part of the year than the grain crops they have no storage space for (Palmer, 1985).

Agricultural extension has also suffered from gender bias. Despite the fact that most of Africa's farmers are female, for example, the vast majority of extension officers are male and trained to deal primarily with men. A 1981 survey found only three percent of all agricultural extension agents in sub-Saharan Africa were female. Not surprisingly, they were paid substantially less than their male colleagues. This pattern, too, is found on every continent; Lynn Bennett of the World Bank notes that agricultural extension services in India largely bypass the 40 percent of the country's farm workers who are female. And studies of extension services in Central and South America find similar biases (Saito and Weidemann, 1990; Bennett, 1989; Poats et al., 1988).

The third trend undermining women's ability to provide food security, beyond the diminishing availability of land and other resources, has been the **technological "modernization" of developing countries**. Mechanization, like privatization, has tended to benefit men who own land, while only making things harder for women who do not. Scattered introduction of tractors and improved animal-powered equipment in Africa, for instance, has lightened the workload of male landowners, enabling them to expand cultivation of their cash crops. But at the same time it has increased the amount of labor done by their wives, who must spend more time doing the "women's work" on the expanded fields (see, for example, Cleaver and Schreiber, 1992).

Aggravating these three pressures on food security is the fact that the labor available to subsistence households in many countries has become increasingly scarce. Greater migration of males to cities, low wages, abandonment, divorce, widowhood, and—and in some cultures—the practice of having more than one wife, all have conspired to reduce the amount of labor and of income contributed to families by men. This, in turn, puts increasing pressure on women to make up the labor shortages by carrying out all the traditionally male activities as well as their own. Under these intensifying time and labor pressures, state Cleaver and Schreiber of the World Bank, "[African] women farmers have little choice but to continue to practice [agricultural methods] such as sharply shortened fallow periods, that are neither environmentally sustainable or viable" over the long term (Cleaver and Schreiber, 1992).

The growing number of female-headed households—now swelling the ranks of the poor in virtually every country of the Third World—is one indication of how widespread these conditions are. Estimates indicate that women are the sole breadwinners in one-fourth to one-third of the world's households, and at least one-fourth of all other households rely on female earnings for more than 50 percent of total income (Agarwal et al., 1990).

■

THE LOSS OF FOREST RESOURCES

Just as they are losing their access to farmland, women in subsistence economies are losing their access to forest resources in every region for most of the same reasons: privatization, lack of credit and extension services, and male veto power over women's decisions.

In many countries, large areas of communal forest land have been privatized and set aside for agriculture, resulting in widespread deforestation and a decline in women's access to woodland resources. On the island of Zanzibar, for example, commercial clove tree plantations began to replace natural forests in the late 19th century. One hundred years later, the spread of commercial agriculture is creating a fuel crisis within subsistence households, which now spend up to 40 percent of their income on fuel. Similarly, in western Kenya communal resources have declined in areas where increases in private land ownership have "commercialized" trees (Williams, 1991b).

In India, much of the commons land now disappearing into government and private hands was previously used by village women to secure fuel according to community rules. "Contrary to the popular belief that it is the gathering of wood for fuel that is [primarily] responsible for deforestation and fuelwood shortages," states Bina Agarwal, "existing evidence [in India] points to past and ongoing state policies and schemes as significant causes." In a seven-state survey, village commons areas were shown to have declined 26 to 58 percent as a result of land grabs by commercial interests and large landowners. Agarwal contends that, if sustained, the widespread appropriation of sacred groves and other communal land by government and private interests for cash crops, dams, and commercial timber will have claimed what remains of India's forests in 45 years (Agarwal, 1991).

Within the subsistence economy, differences in the access of men and women to trees are very similar to, and usually linked with, those of land for food crops. In one region of Kenya where women cannot own land, for example, they are also restricted from planting trees. According to custom, control over land is determined by the ownership or planting of trees on it. Not surprisingly, men in the area have opposed women's attempts to increase supplies of biomass. In northern Cameroon, some men only allow their wives to plant papaya trees, which are short-lived and do not confer land rights (Molnar and Schreiber, 1989).

Increasingly, wood and other biomass products are becoming "cash crops" to which men and women have differential access. Both men and women may be interested in raising cash from the harvesting and sale of trees in their control. However, studies show that women usually seek to gather wood and other biomass resources on a sustainable basis, balancing their needs for cash with their need for other products and with the ecological services forests provide. Men, lacking the responsibilities for collection of fuel, fodder, and the like, may have different priorities. Yet it is plantation or farm forestry that receives by far the greatest amount of support from governments and donor agencies (Molnar and Schreiber, 1989; Williams, 1991b; Agarwal and Narain, 1990).

The spread of cash crops has also diminished women's access to biomass resources. In the past, it was common practice for Indian landowners who regularly employed women agricultural laborers to allow them to collect from the land, as part of their "wages," crop residues, grasses, and other biomass for home use. But these sources, too, have dwindled. New crops and harvesting practices leave fewer residues, and thus the total amount of biomass available from lands under cash crops has decreased. Landlords themselves are now interested in the increasing value of biomass for sale in the market. The spread of dairy cooperatives in India, for instance, has created a market for grasses on which to feed milk cows, removing a main source of fodder for village women. "Now I have to steal the grass for my buffalo," states a female agricultural worker from Sikandernagar, "and when the landlord catches me he beats me" (Mies, 1986).

Shifting access to this land away from the poor makes it increasingly difficult for women to procure fuel and other products, leading to their further impoverishment. Time allocation studies show that as a result of scarcity, women are spending more hours in such tasks as the collection of fuel, fodder, and water. In severely deforested areas of India, the typical rural woman and her children now spend four to five hours a day to gather enough fuel for the evening meal. For women in Sudan, the average time required to collect a week's worth of fuelwood has quadrupled since the 1970s (Agarwal, 1988; Cleaver and Schreiber, 1992; Buvinic and Mehra, 1990).

Because of their prominent role as users and managers, including women in the management of forest ecosystems is vital to achieving increases in rural productivity. Still, states Paula Williams, a Forest and Society Fellow with the Institute for Current World Affairs, "most forest policies and most foresters continue to overlook or ignore this." The exclusion of women's needs and expertise has grave implications for the future of forest resources (Williams, 1991b).

In forest management, as in agricultural development, the international community has been of no help in prodding governments to recognize and support the needs of women in subsistence economies. Here, too, women remain curiously invisible to the development community. World Bank consultant Ravinder Kaur concludes that "the importance of other forest products to women and the very active role that women play in forest resource management have remained largely unrecognized and unspecified" (Kaur, 1990).

Tree planting campaigns and international investments to stem deforestation have all but ignored women: Out of 22 social forestry projects appraised by the World Bank from 1984 through 1987, only one mentioned women as a project beneficiary, and only four of 33 integrated rural development programs that involved forestry funded by the Bank over the same period included women in some way (Molnar and Schreiber, 1989).

Williams points out that no African women participated in the World Forestry Congresses of 1978 and 1985. Most national-level forestry plans do not consider women and children as users of these resources, nor how they fare when biomass becomes scarce. Major forestry policies, such as the Tropical Forestry Action Plan prepared by the World Resources Institute in conjunction with multilateral institutions and individual governments, scarcely consider the role of women in forest use and conservation (Williams, 1991b).

Undervaluing women's social and economic contributions hampers efforts to achieve other broad social and environmental goals, such as preserving biodiversity and protecting the role played by forests in water cycles. Women's experience with forest products represents a vast data base on the species scientists regularly lament being unable to catalogue. Tribal women in India, for example, have been found to know medicinal uses for some 300 forest species. "Many people believe," Williams asserts, "that first we should save the world's tropical forests: then we can worry about women and children. Unless we work with women and children, however, it will be impossible to 'save' the humid and dry-land tropical forests. You cannot save the trees when you ignore over half the users and managers of forest resources" (Abramovitz and Nichols, 1992; Williams, 1991b).

■

FEMALE POVERTY AND THE POPULATION TRAP

From food production to control over income, indications are that the position of women within subsistence economies is growing increasingly insecure. As women's access to resources continues to dwindle in subsistence economies, their responsibilities—and the demands on their time and physical energy—increase. They are less likely to see the utility of having fewer children, even though population densities in the little land left for subsistence families are rapidly increasing.

These trends extend from rural areas into urban ones. Environmental degradation and impoverishment have driven millions of people into the slums and shantytowns of Third World cities. In these **urban subsistence economies**, women maintain their heavy burden of labor and responsibility for the production of subsistence goods. Moreover, urban women are also discriminated against in the access to resources they require to support their families. "When urban authorities refuse to provide water supply, sanitation, and refuse collection to low-income urban areas," write Diana Lee-Smith and Catalina Hinchey Trujillo of the Women's Shelter Network, "it is the women who have to make up for the lack of such services . . . who have to work out ways of finding and transporting water and fuel and keeping their homes reasonably clean, [all] with inadequate support from urban laws and institutions which usually completely fail to comprehend their situation" (Lee-Smith and Trujillo, 1992).

The growing time constraints imposed on women by the longer hours they must work to make ends meet simultaneously lower women's status and keep birth rates high. When they can no longer increase their own labor burdens, women lean more heavily on the contributions of their children—especially girls. In fact, the increasing tendency in many areas of keeping girls out of school to help with their mothers' work virtually ensures that another generation of females will grow up with poorer prospects than their brothers. In Africa, for example, "more and more girls are dropping out of both primary and secondary school or just missing school altogether due to increasing poverty," states Phoebe Asiyo of the United Nations Fund for Women (Asiyo, 1991).

Rapid population growth within subsistence economies, in turn, compounds the environmental degradation—the unsustainable escalation of soil erosion, depletion, and deforestation—first put in motion by the increasing separation of poor farmers from the assets that once sustained them. The health of women and girls, most affected by environmental degradation because of the roles they play, declines further. The cycle accelerates.

This is the population trap: Many of the policies and programs carried out in the name of development actually increase women's dependence on children as a source of status and security. Moreover, environmental degradation triggered by misguided government policies is itself causing rapid population growth, in part as a result of women's economically rational response to increasing demands on their time caused by resource scarcity. Unless governments move quickly to change the conditions confronting women in subsistence economies, rapid population growth will continue unabated.

The objective of reducing population growth is critical to reversing the deterioration of both human and environmental health. However, the myopic divorcing of demographic goals from other development efforts has serious human rights implications for the hundreds of millions of women who lack access to adequate nutrition, education, legal rights, income-earning opportunities, and the promise of increasing personal autonomy.

TOWARD A NEW FRAMEWORK FOR DEVELOPMENT

In the post–Earth Summit era, sustainable development has become a slogan of governments everywhere. Yet given the abysmal record of conventional development strategies in the realms of equity, poverty, and the environment, it is imperative to ask: Development *for whom*? With input *from whom*?

Failing to ask these questions is a failure in the fundamental purpose of development itself. If women in subsistence economies are the major suppliers of food, fuel, and water for their families, and yet their access to productive resources is declining, then more people will suffer from hunger, malnutrition, illness, and loss of productivity. If women have learned ecologically sustainable methods of agriculture and acquired extensive knowledge about genetic diversity—as millions have—yet are denied partnership in development, then this wisdom will be lost.

Without addressing issues of equity and justice, then, development goals that are ostensibly universal—such as the alleviation of poverty, the protection of ecosystems, and the creation of a balance between human activities and the environmental resources—simply cannot be achieved.

In short, development strategies that limit the ability of women to achieve their real human potential are also strategies that limit the potential of communities and nations. Only when such strategies recognize and are geared toward reducing gender bias and its consequences can we begin to solve many of those economic and environmental problems that otherwise promise to spin out of control.

Improving the status of women, and thereby the prospects for humanity, requires a complete reorientation of development efforts away from the current overemphasis on limiting women's reproduction. Instead, the focus needs to be on establishing an environment in which women and men together can prosper. This means creating mainstream development programs that seek to expand women's control over income and household resources, improve their productivity, establish their legal and social rights, and increase the social and economic choices they are able to make.

The first step toward achieving these goals—a step that is consistently overlooked—is to ask women themselves which needs should be accorded top priority. Some answers to the question "What do women want?" were provided in a forum on international health held in June 1991 (see Table 5.2). Among the key needs identified by participants from Africa, Asia, and Latin America were investments in the development and dissemination of appro-

Table 5.2	**Some answers to the question "What do women want?"** (Based on data compiled from a panel discussion at "Women's Health: The Action Agenda for the Nineties," 18th Annual Council on International Health Conference, Arlington, VA, June 23–26, 1991).

- Durable arrangements for the transfer of resources; reductions in (if not cancellations of) the debt burden; direct investments to meet capital requirements.

- Favorable trading terms and better prices for primary commodities such as coffee, tea, and cocoa.

- Access to credit and training; programs for awareness and confidence building.

- Small to medium joint ventures to create jobs; continuing investments in sustainable economic growth.

- Investments in the development and dissemination of appropriate technologies to reduce women's work burdens.

- Access to good food, safe water, and education for both girls and boys.

- Sustainable strategies for the use of natural resources.

- Reallocation of financial resources to critical health care needs, including disease control, maternal and child health and family planning, and development of appropriate health systems.

- Cooperation to establish, expand, and strengthen community-based approaches for promotional and educational activities of family planning and family life education.

- Access to information concerning women's bodies.

- The right to choose the number of children born and to plan families without government interference.

- Access to vaccines, medicines, and equipment.

- Universal access to contraceptives for both men and women.

priate technology to reduce women's work burden, and access for women to credit and training programs.

The second step is to act immediately to increase the productivity of subsistence producers, whether in rural or urban areas. Quick gains can be realized by increasing women's access to land, credit, and the tools and technologies to increase their own and their families' welfare.

Land reform and the enforcement of laws guaranteeing gender equity in the distribution of land resources, for example, need to be assigned high priority in every country. Given the intimate connections among women's lack of access to land, their increasing work burden, and their subsequent reliance on children, Third World land distribution and allocation policies should be at the top of the agenda for groups concerned about the environment, human rights, hunger, and population issues. By mounting a concerted campaign, grass roots groups can focus the attention of media, governments, and international agencies on the issue of land reform. Pushing for simultaneous reform of other policies that discriminate against women—such as those limiting access to credit, improved technologies, and farm inputs—is equally important.

The third step is to examine critically the definitions and assumptions made by conventional development policies in order to collect information that creates a real picture of subsistence economies. As the evidence shows, women are responsible for producing an equal or larger share of the goods on which families depend for survival, yet they often are denied credit for their contributions either because it is not in the form of cash income or because it is simply assumed their income is relatively less important than that of men. These assumptions need to be changed.

A redefinition of the concepts of productivity, value, and work to include activities that are indeed productive—such as those that yield family income in goods rather than in cash or that support people without degrading the environment—would dramatically alter the base of relevant information sought by those truly interested in improving the human prospect. This is a necessary precondition to environmentally sound economic systems. As Lee-Smith and Trujillo of the Women's Shelter Network point out, "careful management of the local resource base to provide for continued human sustenance is something women have been doing for a long time . . . [and is] what is required of the whole human community to achieve sustainable development at the planetary level" (Lee-Smith and Trujillo, 1992).

Following from changes in how work is defined is the need to generate critical new information by redirecting some of the research on the benefits of development. Already, the collecting of gender-disaggregated data on a small scale has helped policy makers recognize the disparate effects on men and women of conventional gender-blind development practices. However, for many areas of the economy in which women play important but officially ignored roles, not enough information is available yet to truly inform public policy. Such gender-based data need to be incorporated into all relevant areas of economics, health, and environment.

Research and development in the sciences and in appropriate technologies needs to be far more gender sensitive, not only to benefit women but to benefit *from* them, especially in areas of crop production and biodiversity. Focusing research on the needs of women in subsistence economies would dramatically boost food crop and forest production within a decade.

However, this cannot be achieved unless women enjoy the same degree of independence and freedom of choice as men. Governments and international agencies also need to be pushed to recognize the effects of their policies on how men and women interact, and on how such resources as money, food, and the opportunities for learning are allocated within the household. Instead of increasing the division between men and women, the goal of development should be to seek more cooperation between the sexes in achieving mutual goals of ending hunger and poverty, and securing the environment.

Family welfare cannot be improved without increasing women's access to and control over resources that are essential to improving nutrition, lowering infant mortality, reducing fertility, and changing a wide range of other variables including violence against women. These ends can most easily be achieved in the short run by directing resources into the formal education of young girls and the formal and informal education and training of older women. At the same time, policies that increase women's access to information and training as well as credit will improve their employment prospects and enable women entrepreneurs to establish businesses, earn income, and create jobs.

Experience suggests that winning these reforms will not be easy. Gender bias—like that based on race, class, and ethnicity—dies hard. Much of the information regarding women's roles in agriculture and forestry, for example, has been available to governments and development planners for two decades and has yet to provoke real changes in policy. In part, this is due to the lack of support by grassroots women's groups in the wealthy countries for a broad-based, politically charged international movement to combat discriminatory development policies. Without doubt, many of the cultural and economic obstacles faced by women in countries such as Brazil, India, Thailand, and Zimbabwe are vastly different from those faced by the majority of women in more prosperous countries like the United States and France. In reality, though, many of these differences are matters of mag-

nitude rather than substance. A number of trends, including the growth in, and disproportionate poverty of, female-headed households, are as evident in the urban and rural areas of
industrial countries as they are in the Third World. Moreover, issues such as equal pay for
work of equal value, domestic violence, reproductive health and freedom, and environmental sustainability are universal.

Addressing these issues requires closer cooperation between women's movements in the
world's North and those in the South. Early signs of a truly international women's movement can already be seen, for example in the U.N.-sponsored Global Assembly on Women
and the Environment and the World Women's Congress, both held in Miami, Florida, in
November 1991.

These concerns are not for women only. Indeed, it is in the interest of every person—
from the poor farmers of Sikandernagar to the chiefs of industry, from grassroots activists
to heads of state—to combat gender bias. Ultimately, the changes needed to make women
equal partners in development are the same as those required to sustain life itself. Nothing
could be more important to human development than the reform of policies that suppress
the productive potential of half the Earth's people.

■

REFERENCES

Abramovitz, J., and Nichols, R. 1992. Women and biodiversity: Ancient reality, modern imperative. *Development* no. 2.

Acsadi, G., and Johnson-Acsadi, S. 1987. Safe motherhood in South Asia: Sociocultural and demo-graphic aspects of maternal health. Unpublished background paper prepared for the Safe Mother-hood Conference, Pakistan, 1987.

Agarwal, A., and Narain, S. 1990. *Strategies for the Involvement of the Landless and Women in Afforestation: Five Case Studies from India* (Geneva: International Labour Organization).

Agarwal, B. 1988. Neither sustenance nor sustainability: Agricultural strategies, ecological degradation and Indian women in poverty. In: B. Agarwal (ed.), *Structures of Patriarchy: State, Community, and Household in Modernising Asia* (London: Zed Books).

———. 1991. *Engendering the Environment Debate: Lessons from the Indian Subcontinent.* Distin-guished Speaker Series Paper 8 (East Lansing, MI: Center for Advanced Study of International Development, Michigan State University).

Agarwal, B., et al. 1990. *Engendering Adjustment for the 1990s: Report of a Commonwealth Expert Group on Women and Structural Adjustment* (London: Commonwealth Secretariat).

Asiyo, P. 1991. What we want: Voices from the South. Unpublished paper presented at Women's Health: The Action Agenda for the Nineties, 18th Annual Conference, National Council of International Health, Arlington, VA, June, 1991.

Bennett, L. 1989. Gender and poverty in India: Issues and opportunities concerning women in the Indian economy. Unpublished internal document (Washington, DC: World Bank).

Blumberg, R.L. 1990. *Gender matters: Involving women in development in Latin America and the Caribbean.* Agency for International Development Bureau for Latin America and the Caribbean (Washington, DC: U.S. Agency for International Development).

Buvinic, M., and Mehra, R. 1990. *Women in Agriculture: What Development Can Do* (Washington, DC: International Center for Research on Women).

Carr, M. 1985. Technologies for rural women: Impact and dissemination. In: I. Ahmed (ed.), *Technology and Rural Women: Conceptual and Empirical Issues* (London: Allen & Unwin).

Census Commissioner. 1991. *Census of India. Provisional Population Totals. Paper One of 1991* (New Delhi: Registrar General).

Chatterjee, M. 1991. *Indian Women: Their Health and Productivity* (Washington, DC: World Bank).

Cleaver, K., and Schreiber, G. 1992. *The Population, Agriculture, and Environment Nexus in Sub-Saharan Africa* (Washington, DC: World Bank).

Davidson, J. (ed.). 1988. *Agriculture, Women and Land: The African Experience* (Boulder, CO: Westview Press).

Deere, C.D., and Leon de Leal, M. 1982. *Women in Andean Agriculture.* Women Work and Development Paper 4 (Geneva: International Labour Organization).

Dey, J. 1984. *Women in Food Production and Food Security in Africa.* Women in Agriculture Paper 3 (Rome: U.N. Food and Agriculture Organization).

Durning, A.B. 1989. *Poverty and the Environment: Reversing the Downward Spiral.* Worldwatch Paper 92 (Washington, DC: Worldwatch Institute).

Ghosh, A. 1991. Eighth plan—Challenges and opportunities–XII, health, maternity and child care: Key to restraining population growth. *Economic and Political Weekly,* April 20, 1991.

Gittinger, J.P. 1990. *Household Food Security and the Role of Women.* World Bank Discussion Paper 96 (Washington, DC: World Bank).

ILO (International Labour Organization). 1984. *Rural Development and Women in Africa* (Geneva: International Labour Organization).

Jacobson, J.L. 1988. The forgotten resource. *World Watch.* May/June, 1988.

———. 1991. *Challenge of Survival: Safe Motherhood in the SADCC Region* (New York: Family Care International).

Jodha, N.S. 1990. Depletion of common property resources in India. In: G. McNicoll and M. Cain (eds.), *Rural Development and Population: Institutions and Policy* (New York: Oxford University Press and the Population Council).

Kaur, R. 1990. Women in forestry in India. Unpublished background paper prepared for World Bank review on women and development in India (Washington, DC: World Bank).

Lee-Smith, D., and Trujillo, C.H. 1992. The struggle to legitimize subsistence women and sustainable development. *Environment and Urbanization.* April, 1992.

Mies, M. 1986. *Indian Women in Subsistence and Agricultural Labour.* Women, Work and Development Paper 12 (Geneva: International Labour Organization).

Milimo, M.C. 1987. Women, population, and food in Africa: The Zambian case. *Development: Seeds of Change* 2 (3).

Molnar, A., and Schreiber, G. 1989. *Women and Forestry: Operational Issues.* Women in Development Working Papers (Washington, DC: World Bank).

Palmer, I. 1985. The impact of agricultural development schemes on women's roles in food supply. In: *Femmes et Politiques Alimentaires* (Paris: Editions de L'ORSTOM).

Poats, S.V., et al. 1988. *Gender Issues in Farming Systems Research and Extension* (Boulder, CO: Westview Press).

Prosterman, R.L., and Riedinger, J.M. 1987. *Land Reform and Democratic Development* (Baltimore, MD: Johns Hopkins University Press).

Ram, A. 1991. *Women's Health: The Cost of Development in India.* Rajasthan Status Report (Washington, DC: Panos Institute).

Rocheleau, D. 1988. Women, trees, and tenure: Implications for agroforestry. In L. Fortmann and J.W. Bruce (eds.), *Whose Trees? Proprietary Dimensions of Forestry* (Boulder, CO: Westview Press).

Russo, S., et al. 1989. *Gender Issues in Agriculture and Natural Resource Management* (Washington, DC: U.S. Agency for International Development).

Saito, K., and Weidemann, C.J. 1990. *Agricultural Extension for Women Farmers in Africa.* World Bank Discussion Paper 3 (Washington, DC: World Bank).

Schrijvers, J. 1988. Blueprint for undernourishment: The Mahaweli river development scheme in Sri Lanka. In: B. Agarwal (ed.), *Structures of Patriarchy: State, Community, and Household in Modernising Asia* (London: Zed Books).

Sen, A. 1990. More than 100 million women are missing. *New York Review of Books.* December 20, 1990.

UNDIESA (United Nations Department of International Economic and Social Affairs). 1991. *The World's Women: Trends and Statistics* 1970–1990 (New York: United Nations).

UNDP (United Nations Development Programme). 1991. *Human Development Report 1991* (New York: Oxford University Press).

Williams, P.J. 1991a. Women's participation in forestry activities in Africa: Preliminary findings and issues emerging from case studies. Unpublished paper prepared for the Institute for Current World Affairs, Hanover, NH, October, 1991.

————. 1991b. Women, children, and forest resources in Africa, case studies and issues. Unpublished paper prepared for Women and Children First, Symposium on the Impact of Environmental Degradation and Poverty on Women and Children, U.N. Conference on Environment and Development, Geneva, Switzerland, May, 1991.

CHAPTER 6

POLITICS AND SOCIETY: POLITICAL CHALLENGES OF CONFRONTING POPULATION GROWTH

■

Anthony C. Beilenson*

INTRODUCTION

There can be no doubt that the rapid **growth of the human population** is the number one problem facing our planet—and yet, judging from the attention the subject gets from the policy makers and from the media, you would certainly never suspect it.

The reasons are obvious. We are talking about a continuing, never-ending phenomenon that is a good subject for an occasional op-ed piece or a few minutes of attention on some weekly TV news show, but it is not something that can be reported or commented on, day after day, in the media. It does not engage our attention the way a terrorist act, an armed conflict, or even a famine does.

Yet, ironically, the implications of—and the actual effects of—the rate at which human beings are reproducing are far more ominous and overwhelming than any combination of tragic occurrences we see on the nightly news. We have to find a way, somehow, to overcome the obstacles that keep us from addressing this greatest problem in a concerted and potentially successful way. And we've got to find a way to do that soon.

POPULATION GROWTH RATE

The world's population recently surpassed the 5.6 billion mark, and it is growing by almost 100 million people every year. When I was born, the Earth's population was at two billion. Since then, it has nearly tripled. Moreover, the next billion will be added in less than 11 years. Twenty-four hours from now, there will be 260,000 more people in the world than there are at this moment. Nearly 95 percent of them will be born in developing countries

*U.S. House of Representatives, 2465 Rayburn Building, Washington, DC 20515

that cannot begin to adequately take care of their current populations—for whom there are too few jobs, inadequate schools, inadequate health care, inadequate amounts of food, and, usually, very little, if any, individual freedom.

Future prospects are even more staggering (Figure 6.1). The United Nations' high fertility population projections indicate that even if the total fertility rate drops from the current world average of 3.2 children per woman to stabilize at 2.5 children—quite a significant reduction—world population could still grow to 12.5 billion by the year 2050. And, if effective action is not taken within this decade—as today's 1.6 billion children in the developing world under the age of 15 reach their child-bearing years—the Earth's population could nearly quadruple to over 19 billion people by the end of the next century (United Nations, 1992).

GLOBAL IMPACT OF POPULATION GROWTH

This rapid growth underlies virtually every environmental, developmental, and national security problem facing the world today. The impact of **overpopulation**, combined with **unsustainable patterns of consumption**, is evident in mounting signs of stress on the world's environment. Under conditions of rapid population growth, renewable resources are being used faster than they can be replaced. Food production, for example, lagged behind population growth in 69 out of 102 developing countries for which data are available for the period 1978 to 1989. Furthermore, the burgeoning of the world's population is having an enormous deleterious effect in such other environmental areas as tropical deforestation, erosion of arable land and watersheds, extinction of plant and animal species, and pollution of air, water, and land.

In much of the developing world, high birth rates, caused in great part by the lack of access of women to basic reproductive health services and information, are contributing to intractable poverty, malnutrition, widespread unemployment, urban overcrowding, and the rapid spread of disease. Population growth is outstripping the capacity of many nations to make even modest gains in economic development. In the next 15 years, developing nations will need to create jobs for 700 million new workers, which is more than currently exist in all of the industrialized nations of the world combined.

Everywhere you look, the prospects are staggering (Figure 6.2). Consider, for instance, a nation like Bangladesh. With a population of 125 million (about half that of the entire United States) jammed into an area the size of Wisconsin, Bangladesh would have little hope of climbing out of its desperate state of severe poverty and underdevelopment even if its popu-

Figure 6.1

Alternative futures: Population projections to 2150, as estimated by the United Nations Population Fund.

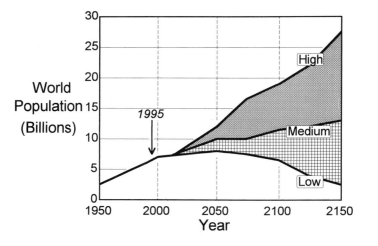

Figure 6.2

Average number of seconds between births by region in 1994, reported by the U.S. Department of Commerce, Bureau of the Census. In Oceania*, a full minute elapses between births.

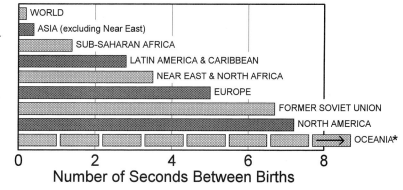

WORLD
ASIA (excluding Near East)
SUB-SAHARAN AFRICA
LATIN AMERICA & CARIBBEAN
NEAR EAST & NORTH AFRICA
EUROPE
FORMER SOVIET UNION
NORTH AMERICA
OCEANIA*

Number of Seconds Between Births

lation remained stable. But it is going to get much worse: In less than 35 years, Bangladesh is projected to add another 100 million people. To illustrate how rapidly the population is growing in just this one country: Bangladesh lost 130,000 people in the great hurricane a few years ago—an immense loss of life. But, then, in just two and half weeks, the population increased by 130,000 as a consequence of more births than deaths.

Bangladesh is only one example. Egypt adds one million people every eight months to a population it already cannot feed. The turbulent Gaza Strip possesses the world's fastest rate of annual population growth, and in Iraq, which comes in a close second, the average woman bears over seven children in her lifetime. Iran's population of 56 million will swell to 130 million in just 30 years. Every day in India 50,000 people are born—and India's population has grown by almost 25 percent over just the last decade, to 850 million people. Every three weeks the population of Africa increases by *one million* people. Nigeria, according to the latest U.N. *medium* projection for 2025, will grow from 108 million to 281 million; Egypt from 52 to 90 million; Ethiopia from 49 to 126 million; Iraq from 19 to 50 million; India from 850 to 1,440 million; Brazil from 150 to 246 million; Kenya from 24 to 79 million; and Mexico from 90 to 150 million. Current U.N. projections for the year 2025 show that Nigeria will have more people than the United States; Iran almost as many as Japan; and Ethiopia nearly twice as many people as France.

And on, and on, and on. Every impoverished, hopeless, and desperate country in the world will see its population double, or more, in the next 30 to 35 years.

To be blunt about it, as Richard Gardner has recently written, nobody "has the slightest idea of how to provide adequate food, housing, health care, education and gainful employment to such exploding numbers of people, especially as they crowd into the mega-cities of the Third World like Mexico City, Cairo and Calcutta.

"The growing numbers of desperate poor will only accelerate the ferocious assault on the world's environment now under way in Africa, Asia and Latin America. . . . Can anyone doubt that even if these medium growth figures are realized, our children and grandchildren will witness unprecedented misery, worldwide violence, and a tidal wave of unwanted immigration throughout the world?"

DOMESTIC IMPACT OF OVERPOPULATION

Overpopulation, however, is not a problem for lesser developed countries only. Rapid population growth in already overcrowded and underdeveloped areas of the world has given rise to an unprecedented pressure to migrate, as workers seek decent and more hopeful lives for themselves and their families. According to a recent report by the United Nations Population Fund (UNFPA), over one hundred million people, or nearly two percent of the world's population, are international migrants, and countless others are refugees within their own countries (United Nations, 1993). Many of the world's industrialized nations are now straining to absorb huge numbers of people, and in the future, as shortages of jobs and living space in urban areas, and resources such as water, agricultural land, and new places to dispose of waste grow even more acute, there will be even greater pressure to emigrate.

In Los Angeles, we observe the effects of these emigration pressures. Communities in Los Angeles County, where enormous numbers of both legal and illegal immigrants are settling, are literally being overwhelmed by the burden of providing educational, health, and social services for the newcomers. Moreover, the problem will get bigger: *Largely because* of immigration, California's population is expected to grow from 31 million, where it stood in 1990, to 63 million by the year 2020.

■

WHAT ARE THE SOLUTIONS?

We *know* what is required to defuse the population explosion: More economic development in the developing world; better education and employment opportunities for women; and universal access to affordable, quality family planning services. A recent Demographic and Health Survey (DHS) study indicates that in most developing countries more than half of the married women do not want any more children; tens of millions more would like to delay subsequent births (Westoff and Ochoas, 1991). However, at least 120 million married women are not using contraception, largely due to lack of availability, even though they wish to avoid pregnancy (Figure 6.3).

The hopeful news is that family planning programs have been remarkably successful worldwide. In general, average fertility falls by about one birth for every 15 percentage-point increase in the number of married couples using contraception. Since the early 1960s, contraceptive use worldwide has gone up from roughly 10 percent of couples to over 50 percent today. Over the same period, the number of births per woman dropped from 6 to 3.3, almost half the fertility of just one generation ago.

The effects of these efforts are already apparent. A study in the 1992 *State of World Population* published by the UNFPA, which examines the linkages between population growth rates and economic development, shows that those countries that took early and effective action to slow population growth in the early 1960s and 1970s did significantly better in the economically difficult years of the 1980s. In fact, countries with slower population growth saw their average incomes per person grow 2.5 percent a year faster than those with more rapid population growth.

Education and access to contraception also has a positive effect on both infant and women's mortality rates. Worldwide, the combination of better birth spacing and the elimination of births to adolescents could avoid at least three million infant deaths a year, or 20 percent of the estimated 15 million deaths a year to children under five. Moreover, adequate family planning would reduce the enormous number of deaths from pregnancy-related problems, which the World Health Organization estimates to be the cause of between 20 percent and 45 percent of all deaths among women ages 15 to 49 in the developing world (World Health Organization, 1986).

Time is of the essence. How quickly we provide worldwide access to family planning is crucial. Like compound interest applied to financial savings, high fertility rates produce ever-growing future populations. Two examples: If a woman bears three children instead of six, and her children and grandchildren do likewise, she will have 27 great-grandchildren rather than 216. If Nigeria, which now has 109 million people, reaches replacement fertility by 2010 rather than 2040 (as currently projected), its eventual population would be 341

Figure 6.3

Unmet need for family planning among currently married women for selected countries, 1985 or later, according to Westoff and Ochoas, 1991 (Table 4.2).

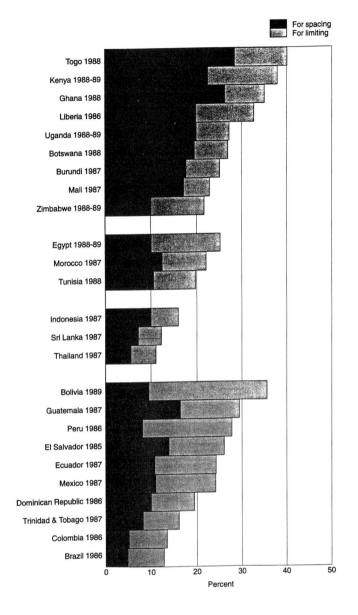

million rather than 617 million. Thus, what we achieve in the way of making family planning services available in this decade will determine whether world population stabilizes at double today's level or at triple that level—or more.

The model for achieving universal access to voluntary family planning by the year 2000, and the stabilization of population at the earliest feasible date, is the **1989 Amsterdam Declaration,** a practical blueprint issued by the 80 governments—including the United States—that participated in the United Nations Amsterdam Forum on Population. That plan is based on studies that indicate if quality contraceptive information and supplies were readily available, about 75 percent of reproductive-age couples in most countries would use them (compared with just over 50 percent today). At the 75 percent level of contraceptive use, people tend to have an average of just over two children per couple, which results in replacement-level fertility.

The Amsterdam plan calls on industrialized countries to dedicate just four percent of their foreign aid to population assistance. At this time, only Norway funds at this level. The United States currently provides about three percent of its foreign aid budget, or $502 million, to international population assistance, but fully funding our share of the Amsterdam plan next year would require raising that amount to $800 million, or to put it another way, the United States could be doing its fair share to make family planning services available universally by devoting a mere .05 percent—five-hundredths of one percent—of total federal expenditures to international family planning programs (Table 6.1).

■

POPULATION POLITICS

For those of us in Congress who care about overpopulation, the past 12 years have been enormously frustrating. American leadership on this issue, traditionally strong, was virtually abandoned by the last two administrations. In the 1980s, the United States stopped funding the two major organizations devoted to population efforts: the International Planned Parenthood Federation and the UNFPA. In a 1984 policy statement, President Reagan rejected the notion of a global population crisis and characterized population growth as a "neutral phenomenon" in the development process.

This opposition to family planning assistance revolved around the politics of abortion. The Reagan administration's so-called "Mexico City Policy" denied U.S. funds for nongovernmental organizations "which actively promote abortion as a method of family planning in other nations," and the UNFPA was refused funding on similar grounds. Ironically, these policies, aimed at stopping abortions, have meant *more* abortions because of curtailed access to sex education and contraceptives.

Fortunately, the **Clinton administration** seems to want to restore U.S. leadership in this area. One of President Clinton's first actions on taking office was to issue an executive

TABLE 6.1

AID obligations for international family planning, 1986–1993, including funding from all sources (Populations Planning Account, Development Fund for Africa, and Economic Support Fund), as compiled by the U.S. Agency for International Development, Office of Population.

Source	Year							
	1986	1987	1988	1989	1990	1991	1992	1993
	Funding (millions of $)							
Office of Population	133.5	114.9	111.0	111.0	142.4*	208.0*	167.1*	240.8
Bilateral and Regional Programs								
Asia	59.8	39.9	44.8	54.6	48.8	24.3	51.8	56.7
Africa	41.4	38.1	32.0	37.9	35.8	57.1	54.9	76.2
Latin America/Caribbean	30.2	24.6	24.1	29.2	26.5	21.8	32.3	33.2
Europe/Near East	29.8	29.1	31.6	19.9	22.0	37.1	17.2	8.3
UNFPA	0.0	0.0	0.0	0.0	0.0	0.0	0.0	14.5
Others	0.9	4.8	1.1	2.0	4.4	2.8	3.1	4.5
Total All Sources	295.6	251.4	244.6	254.6	279.9	351.0	326.4	434.2

*1990–1992 Office of Population levels include budget transfers from AID missions for contraceptives (1990 = $21.7 million; 1991 = $26.8 million; 1992 = $10.1 million).

order lifting the U.S. restriction on funding UNFPA and repealing the Mexico City Policy. In his remarks, President Clinton explained that this step "will reverse a policy that has seriously undermined much needed efforts to promote safe and effective family planning programs abroad, and will allow us to once again provide leadership in helping stabilize world population." The president characterized this action as "one of the most significant environmental steps the United States can take."

In addition to these statements of support, there are other indications that the Clinton administration will actively support population assistance. This year, the president requested an additional $101 million in funding for international family planning assistance, which represents an 18 percent increase over last year's amount and was the most significant increase among all programs addressing global concerns. The Clinton administration has also indicated that it intends to overhaul United States foreign aid programs, moving away from the current system of country-by-country funding, to supporting a few broad national goals, among them "global environment, health and population."

In addition, President Clinton has appointed several strong proponents of family planning to key positions, including Brian Attwood as director of the United States Agency for International Development and former Senator Tim Wirth, a long-time supporter of population programs, as undersecretary of state for global affairs. Mr. Wirth's entire job, in effect, will be to provide leadership for these very issues, and at a recent international conference in New York, he articulated the United States' commitment to family planning, voicing strong support for reproductive choice and access to safe abortion. He emphasized three major concerns: women's health and status, population and the environment, and migration, and he made a point of rejecting the Reagan administration's position that population growth is a "neutral phenomenon" for the environment and development.

In Congress, those of us who feel strongly about these issues have a caucus through which we work to promote international population programs. The **Congressional Coalition on Population and Development** is a bipartisan group, which I currently cochair, along with Congresswoman Constance Morella of Maryland. Founded in 1985, the coalition mainly educates other members of Congress about population issues, and it also plans legislation and coordinates political support for family planning initiatives.

One of the most important efforts of the coalition has been to increase funding for family planning assistance. For the past several years, I have been the principal sponsor of the congressional request for such increased funding to the House Appropriations Foreign Operations Subcommittee, which is largely responsible for determining the level of funding for most foreign assistance programs. In our request, which in 1993 was signed by 156 of my colleagues in the House of Representatives, we have urged the subcommittee to provide funding for international family planning programs at a level consistent with the recommendations of the Amsterdam Declaration. Our efforts have helped raise the U.S. contribution to population assistance from $270 million in 1990 to $507 million this year— still not our full obligation under the Amsterdam plan, but a remarkable increase in light of the tremendous pressure to cut spending, especially for foreign aid, during that period.

Even though additional funding is important, however, what is really needed is a comprehensive approach to address overpopulation. This is why Congresswoman Morella and I, along with Senators Jeff Bingaman and Alan Simpson, have introduced the **International Population Stabilization and Reproductive Health Act**, which would make the goal of population stabilization, along with the improvement of women's and children's health, a primary purpose of U.S. foreign policy. Many groups and individuals with broad experience in population matters spent a great deal of time and effort helping us develop this legislation; the result, we believe, is a truly comprehensive and workable approach to population stabilization and reproductive health.

Our bill seeks to focus U.S. foreign policy on a coordinated strategy that will bring about the widespread availability of contraceptive services and health programs, as well as educational, economic, social, and political opportunities necessary to enhance the status of women. It sets specific health objectives, program descriptions, and funding targets to guide U.S. population programs, and expands U.S. efforts for the treatment and prevention of AIDS and other sexually transmitted diseases. The funding levels in the bill are consistent with the recommendations of the Amsterdam Declaration.

This legislation would increase U.S. commitment to providing universal access to basic education, with an emphasis on eliminating the gap between female and male literacy levels and school enrollment, and on promoting equal opportunities for women. Initiatives to increase infant and child survival, as well as to ensure the health and safety of pregnant women, are included as a critical component to achieving the bill's goals. In addition, our bill urges the multilateral development banks to increase their support for population activities to at least $1 billion by the end of 1999; expresses support for the United Nations Forward Looking Strategies for the Advancement of Women, as adopted in 1985 by the United Nations Conference ending the Decade for Women; and recommends Senate ratification of the United Nations Convention on the Elimination of All Forms of Discrimination against Women, which was signed by the United States in 1980.

The House Foreign Affairs Committee recently held a hearing on population assistance, where I testified in support of our legislation, and in 1994, as the committee and the Clinton administration began rewriting the 1961 Foreign Assistance Act, I am hopeful that our bill will be incorporated as a key component of this foreign aid reform.

POLITICAL PROBLEMS IN ADDRESSING POPULATION GROWTH

Although many of my colleagues in Congress seem to agree about the importance of family planning programs, there are a number of obstacles to providing more assistance in this area. The most difficult problem we face in trying to address population growth is that it is hard to focus attention on the issue when there are other, more immediate issues demanding our attention.

It is encouraging that President Clinton and members of his administration understand the importance of taking action to deal with population growth. Nevertheless, this issue has obviously not been a priority. They have been, of necessity, focused on other matters—the economy and the federal budget deficit; health care reform; NAFTA; and, in foreign policy, crises in Somalia, the former Soviet Union, Haiti, the Balkans, and elsewhere.

Yet even if the administration were to make international family planning a high priority, we would still face a major obstacle to providing adequate funding for it because of our **federal deficit problem**. Foreign aid, in general, is probably the least popular area of federal spending—in fact, the only area our constituents want us to cut more than foreign aid is Congress. Foreign aid is a regular target of efforts to cut the budget, and it has been held at a relatively low level for the last several years.

In the budget bill passed in August 1993, we froze for five years all discretionary spending—that is, foreign assistance, defense, and all the domestic programs that are not entitlements. Thus, we have pitted international spending against domestic spending, where new administration initiatives will demand substantial increases, and against defense, where we hope to bring down spending substantially but cannot be sure we will.

This means that there will be tremendous pressure to cut the total level of spending for foreign aid even as the demands for those dollars will be increasing. The breakup of the Soviet Union and the Warsaw Pact, the recent success of the Middle East peace process, elections in South Africa, and other significant international developments necessitate increased, or at least constant, contributions to these regions. These demands are likely to take precedence over population, education, and hunger programs.

Besides the budget problems, we have to contend with objections to providing population assistance for various **ideological reasons.** First, there is a theory that rapid population growth in developing countries helps, not hinders, economic and technological development, thus alleviating poverty and other economic and social ills. Although this thinking is derided by virtually all experts who deal with this issue in one way or another, it seems to be getting a disproportionate amount of ink on the op-ed pages of our newspapers.

Julian Simon, professor of management at the University of Maryland, who seems to be the leading spokesman for this school of thought, has suggested that the world is underpopulated and that every country in the world could sustain a Western standard of living. He and other proponents point to the concurrent growth in population and development in Western Europe from 1650 onward as the basis of their contention. This theory, however, ignores the preponderance of evidence as to the effects of overpopulation and consumption patterns on the environment and on living standards. Moreover, the historical comparison between population growth in Western Europe and that of the developing world is grossly misleading. Over the past 300 years, Western Europe's population has grown from 105 million people to around 502 million today. The developing world, however, is adding over 90 million people to its population each year, and many countries, such as Iran and Zaire, are likely to *double* their populations in about 20 years. This rapid rate of growth, not merely growth itself, is crippling the ability of governments to provide for their citizens, consuming resources faster than they can be replaced, and undermining long-term economic development (Figure 6.4; Table 6.2).

The danger of the pro-population-growth movement is not that lots of people will adopt that point of view—after all, it defies common sense—but that it will raise questions in peoples' minds about whether there is agreement among experts that there is an overpopulation problem. As with any issue, unless there is a consensus that there is in fact a problem, it is difficult to get political action to address it.

There are others who do not dispute a need for family planning assistance in developing nations but question why the United States should provide it, or why we should increase our commitment. When I testified before the House Foreign Affairs Committee recently on our population stabilization bill, I was asked by one of my more conservative colleagues why we should increase family planning assistance to other countries when those countries are spending *their* money on weapons, and by one of my more liberal colleagues why we spend

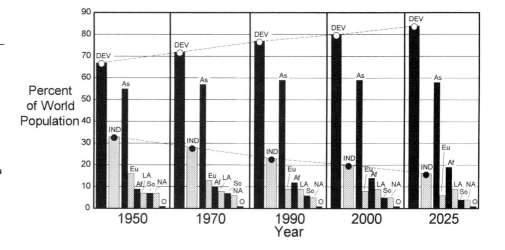

Figure 6.4

Estimated and projected population by percent of world population by region, 1950–2025, based on data compiled by the United Nations Population Division (United Nations, 1991). The percent of the world population residing in developing countries (DEV) is projected to increase, whereas that in industrialized countries (IND) is projected to decrease. As, Asia (largely China and India); Eu, Europe; Af, Africa; LA, Latin America; So, the former Soviet Union; NA, North America; O, Oceania.

Table 6.2	**Estimated and projected population by region,** 1950–2025, based on data compiled by the United Nations Population Division (United Nations, 1991).

Region	Year				
	1950	**1970**	**1990**	**2000**	**2025**
	Population (millions)				
WORLD TOTAL	2,516	3,698	5,292	6,261	8,504
Developing Countries	1,684	2,649	4,086	4,997	7,150
Industrialized Countries	832	1,049	1,207	1,264	1,354
Asia	1,377	2,102	3,113	3,713	4,912
Europe	393	460	498	510	515
Africa	222	362	642	867	1,597
Latin America	166	286	448	538	757
Former Soviet Union	180	243	289	308	352
North America	166	226	276	295	332
Oceania	13	19	26	30	38

the money abroad when we need more family planning assistance here in the United States, especially in our inner cities. Both of these points of view have a lot of populist appeal and have to be answered in a strong and persuasive way if we are to counteract them.

Opposition to abortion also remains a problem. Over the years, Congress has taken several actions to restrict U.S. involvement in what has been deemed by some critics as undesirable family planning activities abroad. The 1973 "Helms Amendment" to the Foreign Assistance Act prohibited the use of foreign assistance funds to pay for abortions; in 1981 Congress added a provision prohibiting funds for biomedical research relating to abortions; and the policies of the Reagan and Bush administrations terminated funding for any organization (but not governments) that was involved in voluntary abortion activities. Other than the change in the Mexico City Policy, the anti-abortion lobby has managed to hold ground on prohibiting U.S. involvement in anything that could be construed as support for abortion overseas, even though common sense dictates that the availability of safe and effective family planning methods and services can help prevent millions of women from having abortions.

Finally, some people are reluctant to support family planning assistance because of a fear that governments or others will use family planning in coercive ways. This seems to be more of an issue as we develop more effective and long-lasting forms of contraception. Nevertheless, the more effective our contraceptives, the more freedom people have to choose the number of children they want and to space them as they wish.

The availability of Norplant, a very effective, easy-to-use new form of contraception, should have been a very welcome development—and I think most reasonable people view it that way. However, Norplant has been vilified by some extreme women's groups, and some extreme minority groups, as something that will be used by authorities to prevent women from having the children they want to have. In Virginia, when the state legislature included some funding for Norplant in its budget, a state representative called it a thinly veiled attempt at "ethnic cleansing" because minority teens would be the likely recipients.

Yet from what we know about teenage pregnancy, rarely do the girls plan to get pregnant and, even if they do, frankly, we *ought* to be pointing out the advantages of delaying childbirth, and providing them effective means for doing so.

This attitude carries over to foreign assistance as well. People with this mindset worry that the powerful, rich, developed nations are using family planning assistance to impose their will on people in weak, impoverished nations. The truth is, however, that nothing could be more coercive, more paternalistic, than denying women access to the means to control their reproduction.

■

CONCLUSION

Let me summarize my thoughts on the political situation we face in trying to address the world population problem. For the first time in many years, we have a president who understands the need for U.S. leadership on the population problem and who has appointed people who share that view to positions where they can be influential in formulating U.S. policy in this area. That in itself is vitally important; without support from the president, there is no hope for curbing world population growth. That was the situation for 12 years and, thankfully, it is over.

But just having a president who cares about the problem is not enough. He has to make it a priority; he has to repeatedly call attention to it by showing how it relates to all the other problems in the world—and how it will affect us here at home if we do not address it. Moreover, members of Congress, especially those who serve on committees dealing with foreign aid, need to do the same. However, neither is likely to happen unless and until world population growth becomes a much bigger issue in the minds of American voters.

That brings me back to my original point: The primary obstacle to addressing world population growth is the lack of attention to this problem; this issue must compete with apparently more pressing problems, most of which have far more direct and immediate impact on regular Americans than does population growth.

I am always encouraged when I talk about the population problem at gatherings with constituents. When one explains how serious and threatening this growth is, people respond very positively to the idea of spending more money on population assistance, and spending it *now*, despite our budget problems, despite their distaste for foreign aid, and even despite any qualms they might have about abortion. Yet how many congresspersons talk to their constituents about world population growth? How many articles do you read in the newspaper about it? How often is it mentioned on the nightly news? How often do you hear the subject discussed on talk radio?

This is *the issue* that all of us on this planet face together. If we don't solve it, none of our other efforts to promote peace, security, and the well-being of people in this country and around the world, now and in the future, will matter. That is the message people need to hear, over and over, to make it possible, politically, finally to take determined action on this most serious problem before it is too late.

■

REFERENCES

United Nations. 1991. *World Population Prospects 1990* (New York: United Nations).

———. 1992. *Long-range World Population Projections: Two Centuries of Population Growth 1950–2150* (New York: United Nations), p. 45.

————. 1993. *The State of the World Population* (New York: United Nations), p. 16.

Westoff, C.F., and Ochoas, L.H. 1991. Unmet Need and the Demand for Family Planning. In: *Demographic and Health Surveys Comparative Studies #5* (Columbia, Md.: Institute for Resource Development/Macro International), p. 37.

World Health Organization. 1986. *World Mortality Rates: A Tabulation of Available Information* (Geneva: WHO).

CHAPTER 7

GLOBAL ENVIRONMENTAL ENGINEERING: PROSPECTS AND PITFALLS

■

Richard P. Turco*

■

INTRODUCTION

The explosion in human population has created daunting environmental problems on a global scale that must be addressed if civilization is to advance and people everywhere are to achieve an acceptable standard of living and comfort. Many of the problems are associated with the widespread application of technology, particularly for the production of energy. In the long run, of course, alternative technologies must be developed, but in the meantime, what can be done to preserve a decent quality of life?

In this chapter we consider the emerging issue of **global environmental engineering (GEE)**, a field that seeks technological solutions to otherwise intractable environmental problems. We examine the prospects for applying GEE to "solve," or at least ameliorate, two of the most dangerous global environmental problems today: global warming and the depletion of the stratospheric ozone. Both of these environmental crises are man-made. Global warming is caused by the accumulation in the atmosphere of greenhouse gases such as carbon dioxide (CO_2), produced primarily from the burning of vegetation and fossil fuels. Stratospheric ozone depletion is caused by the accumulation in the atmosphere of chlorofluorocarbons (CFCs), a family of nearly inert chemicals that have been used worldwide for a variety of purposes including refrigeration. These two crises are among the major global environmental issues of this century.

Two broad approaches to environmental problems can be distinguished: **corrective technologies**, which counteract the polluting side effects of dirty technologies while allowing them to remain in use, and **alternative technologies**, which are clean substitutes for polluting technologies. GEE is mostly concerned with corrective technologies; some visionaries see it as the next logical step in the coevolution of human intelligence and technology. We argue, though, that the proposed technologies are, for the most part, dangerous and that prospects for success with GEE are exceedingly slim.

---

*Department of Atmospheric Sciences, University of California, Los Angeles, CA 90095-1565. This chapter is adapted from Turco, R. 1994. *Earth under Siege: Air Pollution and Global Change* (Oxford, UK: Oxford University Press), chapter 14.

WHAT IS GLOBAL ENVIRONMENTAL ENGINEERING?

According to the vision of GEE, it might sometimes be easier to tinker with the environment to compensate for harmful effects of technology rather than to modify or discontinue the offensive actions. For example, suppose an antidote for the effects of **chlorofluorocarbons (CFCs)** on stratospheric ozone could be found. It might then be easier and cheaper to continue manufacturing refrigerators and air conditioners to use CFCs rather than to redesign them to run on more expensive but less threatening compounds. If the antidote did not work and the ozone layer were eventually depleted, allowing a flood of deadly ultraviolet (UV) sunlight to reach the ground, we could perhaps respond by breeding or genetically engineering new crops to survive the increased UV radiation. If new aggressive pests should then emerge to ravage the crops, we could respond again by developing stronger pesticides. If those pesticides should decimate bird populations, we might just accept that fact as the price for human prosperity. Humans seem to have a tendency to rationalize the trading of long-term environmental damage for short-term pleasure or profit.

Living Thermostats—Natural Compensation

The global climate system has a number of feedback mechanisms, involving the oceans and clouds, that help to damp large climatic swings. Although such stabilizing mechanisms are generally present in environmental systems, they are complex and interlocking, and they are not very well understood. In some cases it seems that human activity might be capable of pushing the environmental system out of the zone of stability, with unknown but almost certainly disastrous consequences.

An example of a complex, partially understood natural feedback mechanism that may influence climate is illustrated in Figure 7.1. It involves the compound **dimethyl sulfide (DMS)**, which is produced by phytoplankton in ocean surface waters. DMS leaks from the ocean into the marine atmosphere where it is then oxidized to form sulfates, which condense into an aerosol, that is, a mist of fine particles. It is thought that these aerosols can increase the **albedo** (reflectivity) of marine clouds, causing them to reflect a greater fraction

Figure 7.1

An example of a natural climate control mechanism. A naturally produced organic compound (DMS, dimethyl sulfide) is released by phytoplankton into ocean surface waters from which it diffuses into the atmosphere where it is oxidized to form sulfate aerosols that alter the reflective properties of the marine stratus clouds that modulate the insolation at the ocean surface where the phytoplankton live. The chain of events feeding back to the phytoplankton is impressively complex and poorly understood. (Modified after Charlson et al., 1987.)

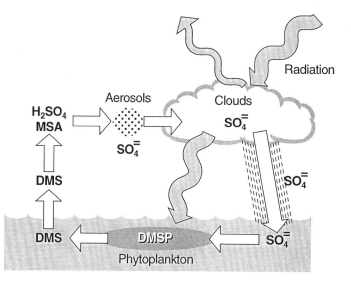

of sunlight back into space, which in turn would tend to cool the Earth's climate. Although the effect of the marine-derived sulfate is still somewhat uncertain, a similar effect is known to cause the unusual behavior of marine clouds following the passage of ships: The fine particle emissions from smokestacks create long-lived "tracks" in the clouds.

Important questions remain unanswered about the plankton-DMS system, however. Is the effect large enough to have an important climatic effect on a global, or even a regional, scale? And is the feedback loop closed, that is, does the climatic change caused by the DMS in turn have a significant effect on the phytoplankton and their rate of production of DMS? Is there a cycle of cause and effect that may either amplify or dampen the climatic effect? The scientific community simply does not yet know the answer to these crucial questions, although it is likely that the DMS-climate connection is very weak.

The Scale of Natural Systems

The DMS-cloud relationship, which represents a rather small part of the global climate system, illustrates the extraordinary complexity of the natural environment. When a new technology inadvertently affects one process in a complex system, the entire system can be thrown into disequilibrium. Thus, any proposed approaches to correcting environmental problems should depend on knowledge of the whole integrated system. All too frequently, however, they do not.

It is relatively easy to dream up schemes for improving the environment by compensating for pollution. The scientific basis for such schemes must be verified, of course, and all side effects—both favorable and unfavorable—must be identified and quantified. However, in addition, engineering practicality and cost must be carefully calculated and assessed, because the enormous scale of the global environment is not often understood by proponents of solutions. To understand the scale of the Earth and its environment, a few statistics will help.

For example, the sun deposits roughly one hundred billion megawatts of power (10^{17} watts) continuously on the Earth. In comparison, a single large power plant generates something like one megawatt of power. Humans, all together, generate about 10 million megawatts (10^{13} watts), or 0.01 percent of the solar input. Thus, heat dissipated by civilization currently does not contribute significantly to planetary warming. Roughly 0.1 percent of the absorbed solar energy is converted by plants into chemical energy stored in biomass, which is released when the biomass decomposes or is burned. Hence, to fill all the present energy needs of society from renewable biomass sources alone, about 10 percent of the world's total biological productivity would need to be harnessed for energy.

The atmosphere weighs five quadrillion (5×10^{15}) metric tons; one part per billion by mass amounts to five million metric tons. The ocean weighs 300 times as much as the atmosphere, and contains heat energy roughly equivalent to 500 years of total solar input. The lower atmosphere has a volume of more than five billion cubic kilometers; the stratosphere is four times larger. The surface area of the oceans is more than 300 million square kilometers. The living organisms on our planet weigh almost one trillion tons, about 200 tons for every living person. The ozone layer weighs four billion tons and is continuously renewed, turning over about once a month.

Humans and their most impressive technological devices and structures are puny in comparison to the scales of the natural world. A large truck can carry 10 tons; a jumbo jet, 100 tons; and a large ship, 1,000 tons. It would take all the people presently on Earth one million years to breathe all the air in the atmosphere. Yet, despite our comparatively small scale, humans are able to do great damage to the global landscape by weakening or destroying critical vulnerable links and components in the environmental systems.

Are scientists and engineers smart enough to recognize potentially hazardous technologies? If the past is any indication, the answer is "not always." Even the most specialized and expensive technologies are not immune to engineering flaws; the Challenger space shuttle and the Chernobyl nuclear power plant are classic examples. On the whole, society appears relatively safe; but the greatest threats of technology arise from subtle traps not yet sprung that lie along the path of progress.

Carbon Dioxide

Most of the conveniences enjoyed by modern society today depend on massive production of **fossil fuel energy**. First coal and, later, oil have proved to be indispensable to industry, and early industrialists who profited from their use never questioned their ultimate value to society. Despite serious air and water pollution, fossil fuel exploitation raced ahead full pace, with no suspicion of global effects. Early ideas concerning carbon dioxide and climate were not immediately connected with the need to control fossil fuels. If a scientist had stepped forward at the beginning of the industrial revolution, warning of possible uncertain effects on the future climate of the Earth, he would have been dismissed as an alarmist. No one really wanted to see a potential problem with such a large cash cow.

Unfortunately, the coal and oil that have driven the industrial revolution have also driven much of the degradation of the global environment. Smog and spills have caused havoc in many parts of the world. And carbon dioxide generated by fossil fuel combustion contributes more to **global climate warming** than any other anthropogenic emission.

Some people believe they have discovered a silver lining—in fact, two silver linings—in the smog that accompanies fossil fuel consumption. First, they note that smog can reflect sunlight, and thus help cool the greenhouse warming (the "albedo effect" described below). Second, and somewhat ironically, ozone in low-altitude smog may block some of the UV radiation leaking through a damaged stratospheric ozone shield. Over the last two decades, as stratospheric ozone has declined a few percent on average over the globe, ultraviolet radiation measured at the surface has not increased in response. In some cases, measurements in urban regions actually indicate *lower* UV radiation intensities as a result of the ozone and other absorbing components of smog. Inadvertently, the same smog that chokes us also shades us from UV radiation.

Is it reasonable, therefore, to suffer bad air quality in order to avoid harmful ultraviolet rays? Hardly. Anyone even slightly concerned with public health is convinced both that smog must be minimized and the ozone layer must be protected. There can be no compromise on either issue; the idea that these problems offset one another is nonsense. Worse, reduction in stratospheric ozone can intensify smog. The increased flux of ultraviolet radiation accelerates smog reactions, cooking up more ozone near the ground where it is hazardous.

It has recently been discovered that while the CO_2 in the emissions from fossil fuel and biomass combustion act to warm the planet, the particulates generated by sulfur dioxide (SO_2) emissions from the same sources tend to reflect sunlight and thus to decrease the temperature of the Earth (the albedo effect). Sulfate aerosols and vegetation smoke particles are effective scattering agents that reduce the amount of solar energy absorbed by the planet.

These offsetting side effects of fossil fuel combustion constitute a treacherous sleight-of-hand on the part of nature. The combustion-generated aerosols remain in the atmosphere only as long as the fuels are burned; when the burning stops, they disappear in a matter of weeks. Carbon dioxide, on the other hand, remains in the atmosphere for hundreds of years. It follows that the warming potential of CO_2 can be masked while fuels are burned and its concentration builds up. Eventually, when the fuels run out, the cooling effect of the aerosols dissipates, and the full-blown warming effect of the carbon dioxide may appear suddenly.

These are examples of the intricate relationships among the physical, chemical, and biological effects of large-scale technology that are not widely appreciated by most people or policy makers. Even among scientists the connections may be overlooked because of the natural tendency in academic circles toward narrow specialization. Most physical scientists, for example, are not familiar with biological systems, and most life scientists are only marginally educated about the physical world. Yet technology sets traps that lurk between academic disciplines.

Chlorofluorocarbons

CFCs are a perfect example of the pitfalls of new technologies, particularly chemical compounds intended for widespread use. CFCs were invented in the early 1930s as a safe replacement for the common refrigerants of that time, including toxic ammonia gas. Since CFCs are nontoxic and can even be breathed without harm, they came to be used as a propellant for deodorants, hair sprays, and other personal hygiene products. Moreover, CFCs are so inert chemically that they can be used in a variety of industrial processes requiring a nonreactive buffer gas, for example, in the production of plastic and rubber foams. Because they are relatively cheap and easy to make, the common CFCs have been widely adopted for every possible use.

The environmental problems that eventually surfaced from the widespread use of the miracle CFC compounds are now legendary. Greenhouse warming and the depletion of the ozone layer are the two major global environmental issues of this century. Like a drunk on the morning after a binge, we are still woozy from the effects of CFCs. We must give them up, yet we are not sure how to live without them.

TECHNOLOGICAL CURES

Technology has caused many of today's most serious environmental problems. Can technology provide solutions as well? If ozone depletion causes cancer, medical techniques might be developed to remove cancerous lesions produced by excess ultraviolet radiation. If global warming causes the sea level to rise, levees might be built to contain the ocean. Are these ideas feasible? What other technological patches might be worth pursuing?

Cooling Down the Greenhouse

Now that combustion gases have been implicated as the primary cause of global warming, the threat of global changes in climate associated with greenhouse warming has fostered a cottage industry in technological cures invented by a slew of climate-sensitized entrepreneurs and technologists. After all, a practical scheme might both forestall the climatic chaos that may follow warming *and* turn a handsome profit for the inventor of the scheme. How reasonable are the possible technological cures for global warming? Let us examine some of them.

The Albedo Effect—Smoke and Mirrors

Most of the schemes for curing global warming involve artificially increasing the Earth's albedo, in order to reflect more of the sun's radiation. Albedo is simply the *fraction* of the total solar energy impinging on the Earth that is reflected away by the atmosphere, the clouds, and the surface. Albedo is one of the key factors controlling the basic planetary energy balance and determining the global climate over long time scales. The albedo, in turn, is controlled by a number of parameters, including conditions of the land surface and the cloudiness of the sky. Albedo can also be affected by the particulate loading of the atmosphere; in general the more particulates, the hazier and more reflective is the air. The only exception occurs in circumstances when highly absorbing particles, like soot, are present in large amounts.

There is a substantial body of data illustrating the effect on climate of albedo change arising from volcanic eruptions. The eruption of Mount Tambora (Indonesia) in 1815 provides the most spectacular example in historical times. Unfortunately the Tambora event was not well documented because it occurred (fortunately) in a remote part of the world, and the geophysical data collected at the time were quite crude. Nevertheless, the so-called "year without a summer" in 1816 is associated with Tambora. In that highly unusual year, unseasonal frosts throughout the summer wiped out crops in New England and parts of Europe.

The fact that volcanoes change climate finds support in a number of geological, biological, and historical records. In one particularly important set of data, damage to tree rings indicates years with extreme weather and climate anomalies. The record of frost damage to the ancient bristlecone pines of the southwestern United States is illustrated in Figure 7.2. The striking point of this study, and many other data, is that climate can be manipulated relatively easily within certain modest bounds. The average year-to-year temperature change from natural causes is small—one degree Celsius or less—but the fact of climate response to specific events is clear.

Variations in the magnitude of the effects caused by volcanic eruptions can be attributed to differences in the amount and type of materials emitted by each volcano, the height of the volcanic injections, and their latitude. Relatively small volcanic eruptions do not cause global effects. In the case of Mt. St. Helens (which exploded in Washington state in 1980), a plume of ash spread over the northwestern United States but did not go much further.

However, recent larger eruptions, such as El Chichon in 1982 (Mexico) and Mt. Pinatubo in 1991 (Philippines), have had major global impacts. Both emitted a large amount of sulfur dioxide, along with ash and other debris. Sulfur dioxide injected into the stratosphere is converted into sulfuric acid aerosols, and these spread over the entire globe, causing, among other effects, spectacular purple twilights. In addition, the amount of sunlight reaching the Earth's surface is reduced, and the climate thus cools slightly. Although the extent of cooling is uncertain, a global average temperature decrease of 0.5°C is expected the year following a major volcanic eruption. The climatic anomaly fades as the volcanic aerosols disappear from the stratosphere over a period of several years. That is long enough to allow a solar energy deficit to build up, but not long enough to cause a long-term climate anomaly.

Changes in solar energy input caused by variations in the sun's brightness or the reflectivity of the Earth have similar climatic implications. Indeed, the effect of a change in the planetary albedo on average surface temperatures over a long time period can be estimated using a simple **climate-change equation,** as follows:

$$\frac{\Delta T_s}{T_s} = -\frac{1}{4}\frac{\Delta \alpha_e}{(1-\alpha_e)} \cong -\frac{\Delta \alpha_e}{3} \qquad \text{(Eq. 7.1)}$$

Here, the normal surface temperature, T_s, will decrease ($\Delta T_s < 0$) as the albedo, α_e, increases ($\Delta \alpha_e > 0$). Since the average surface temperature is close to 300 Kelvin, equation 7.1 can also be expressed approximately as:

$$\Delta T_s \cong -100 \times \Delta \alpha_e \qquad \text{(Eq. 7.2)}$$

Hence, a change in the average planetary albedo of 0.01 (that is, a three percent change from the current albedo of about 0.33) can lead to a surface temperature change of about

Figure 7.2

The record of frost damage in the tree rings of bristlecone pine in the southwestern United States (right column) compared with that of major volcanic eruptions in the Northern Hemisphere. Arrows indicate those years when both frost damage and a relatively high volcanic "Dust Veil Index" (DVI, corresponding roughly to the optical depth of the volcanic aerosol layer in the stratosphere) are known to have occurred. The "NO" symbols on the right column (corresponding to the gray DVI peaks in the left column) indicate years in which a notable dust veil appeared but frost damage was minimal. (Data from Lamarche and Hirschboeck, 1984.)

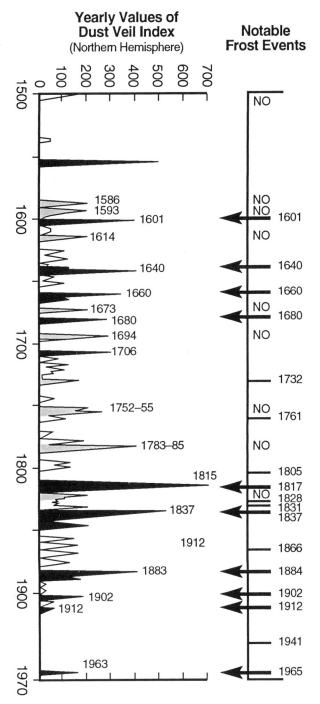

1°C. The effect of any change in albedo would take decades or longer to develop fully, because the ocean heat reservoir is so large that it takes a long time to reach a new equilibrium state. Nevertheless, equations 7.1 and 7.2 are useful for making first-order estimates of global climate changes related to long-term variations in the albedo.

Several possible schemes for purposefully increasing the albedo of the Earth to compensate for an increase in the abundances of carbon dioxide and other greenhouse gases are described below.

Sulfate Aerosols

Whereas the ozone shield protects the Earth from harmful solar ultraviolet radiation, the "sulfate shield," consisting of the aerosols in the lower atmosphere, may help protect the Earth from climate warming. Unfortunately, it is an inefficient prophylactic. Earlier, we discussed one component of these aerosols, those due to dimethyl sulfide (DMS), in the context of a natural climate feedback system over the oceans. Another important sulfate component consists of the particles in polluted air. These aerosols originate from sulfur dioxide, emitted mainly during the combustion of fossil fuels, which undergoes chemical conversion into sulfates, ending up on haze particles, or in acid rain.

Sulfate aerosols are concentrated in the northern midlatitudes over continents. Unlike sulfate particles in the stratosphere, those near the ground in the troposphere have a relatively short residence time in air and are depleted in a matter of hours or days. Aerosols formed over the eastern United States travel over the North Atlantic Ocean but rarely make it as far as Europe.

Besides acting to reduce global temperatures by directly reflecting sunlight, pollution particles, like those generated from DMS, can cause clouds to become more reflective. Both effects are fairly weak, however. The short life of tropospheric aerosols limits their use as climate moderators in environmental engineering schemes. To produce an albedo change sufficient to compensate for greenhouse warming, huge amounts of sulfur would have to be released into the atmosphere, as much as several hundred million tons per year. Filling the air with respirable sulfate particles on such a scale would be generally disastrous to human health and would have a devastating impact on atmospheric visibility.

The role of stratospheric aerosols in controlling the global radiation balance and climate is depicted in Figure 7.3. It happens that the sulfate particles have a much stronger effect on visible solar radiation (at short wavelengths) than on infrared thermal radiation (at long wavelengths). The net effect, therefore, of adding sulfate aerosols (or any other small scattering particles) to the stratosphere is to cool the surface.

Figure 7.4 shows the cooling effect of stratospheric sulfate aerosols. The scattering efficiency of the aerosol layer increases as its opaqueness or **optical depth** increases. Optical depth is a unitless number that measures geometrically the ability of particles to scatter light. The albedo (reflectivity) of the aerosol layer and the cooling effect of the aerosols

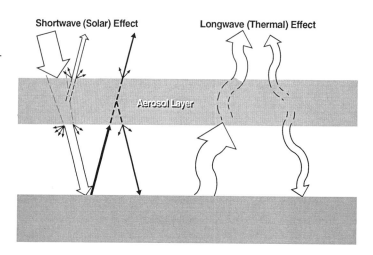

Figure 7.3

Effects of stratospheric sulfate aerosols on the fluxes of solar (shortwave) and thermal (longwave) radiation. Sunlight is affected mainly by aerosol scattering increasing the reflected component, thus reducing the solar radiation reaching the surface. Aerosol absorption of upwelled infrared radiation warms the stratosphere and somewhat enhances the greenhouse effect. The effect of sulfate is typically much larger on solar than on infrared radiation.

Figure 7.4

Surface cooling effect of a layer of stratospheric sulfate aerosols. The optical properties of aerosols are defined in terms of the optical depth of the layer for sunlight in the middle of the visible portion of the spectrum (specifically, at a wavelength of 0.55 microns). The average surface temperature decreases shown in this figure correspond to the new equilibrium conditions of the climate system established after full adjustment to the modified radiation. This new equilibrium state is not reached for a decade or longer, owing to the thermal inertia of the oceans. (Data from Pollack et al., 1976.)

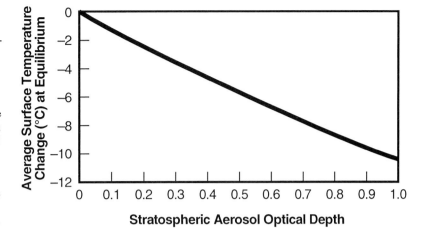

generally increase with increasing optical depth. Over a relatively wide range of values, the relationship between the aerosol optical depth and equilibrium surface temperature change is linear. Thus, if the optical depth of the sulfate layer is doubled, the decrease in surface temperature also doubles. From Figure 7.4 it is apparent that an optical depth of 0.1 can lead to a surface temperature decrease of about 1.5°C.

The optical depth of an aerosol layer varies with wavelength; typically, it decreases roughly in an inverse relationship—that is, if the wavelength doubles, the optical depth is halved. The potential decrease in surface temperature depends on the wavelength dependence of the optical depth, and on the length of time the aerosol layer is present. If the aerosol properties stabilize over a long period, then surface temperature will achieve a new steady state after a decade or so. If the aerosol properties (particle sizes, optical depth, and so forth) vary with time, then the drift in surface temperature will be affected accordingly. Although major volcanic eruptions create optical depths of about 0.1 to 0.2, the cooling is transient and global surface temperature can decrease by only about 0.5°C. Volcanic aerosol layers are too short-lived to achieve a new equilibrium state with maximum surface cooling.

How can the stratospheric sulfate layer be thickened? Lifting 10 to 30 million metric tons of sulfur dioxide to stratospheric heights each year using aircraft has been suggested as a means of mimicking volcanic eruptions. This would require something like a thousand jumbo jet flights every day. The flights would need to continue for as long as the threat of global warming persisted; since carbon dioxide may linger in the atmosphere for a century or more, an artificial aerosol layer would have to be maintained over that span of time. To be effective the sulfur dioxide must be dispersed throughout the stratosphere, and aircraft would have to travel to most points on the globe, flying at all latitudes in both hemispheres. Unfortunately the particle sizes generated in dense sulfur dioxide trails just behind the aircraft would likely be too large for optimal climate modification. Moreover, new planes would need to be designed to fly at much higher altitudes (at least 20 kilometers) than can present jumbo jets, which have a ceiling of about 14 kilometers.

The quantity of aircraft exhaust emitted into the stratosphere during such a program would be unprecedented. Nitrogen oxides and water vapor from the engines could lead to unacceptable ozone depletion through other chemical reactions. Indeed, Congress, in the past, banned stratospheric supersonic aircraft fleets for just this reason.

The Three-Percent Solution

Figure 7.5 depicts an alternative concept to compensate for global warming: The burning of carbon-based fossil fuels that emit trace amounts of sulfur into the atmosphere, primarily in the form of sulfur dioxide. On average, fossil fuels contain a few percent sulfur by mass. Coal, in general, contains the most, oil somewhat less, and natural gas the least. To reduce the problems of sulfurous smog, haze, and acid rain, sulfur is removed from petroleum during processing, or from smokestack effluents of power plants, which can be expensive.

If sulfur could be removed from fossil fuels and somehow used to offset greenhouse warming, the multiple benefits to the environment would appear to be enormous. That concept is sketched in Figure 7.5. **Carbonyl sulfide (COS),** along with hydrogen sulfide (H_2S) and DMS, is among the most important sulfur compounds in nature. COS is produced by bacteria in anaerobic environments and may be absorbed by plants. Combustion processes also generate COS, and it forms as a chemical product when carbon disulfide, CS_2 is photochemically decomposed. Carbonyl sulfide is the most abundant sulfur-bearing gas in the atmosphere, having a relatively uniform mixing ratio of about 0.5 parts per billion by volume throughout the lower atmosphere. Its lifetime in the atmosphere is uncertain but appears to be at least one year, allowing COS to drift far from its sources before it is destroyed. Even when emitted at ground level, COS may travel between the hemispheres. Because of this dilution COS does not generate local sulfate haze, nor does it significantly acidify precipitation on regional scales. Thus, the argument goes, if we could convert the sulfur in fossil fuels to COS, it would largely eliminate local and regional environmental impacts of sulfur emissions.

What about global side effects of the COS emissions? The total amount of sulfur emitted by fossil fuel combustion worldwide approaches 100 million metric tons annually. However, that quantity, after conversion to sulfuric acid, could only marginally increase the acidity of rainfall around the world. In fact, by spreading the sulfuric acid production globally, widespread environmental damage is avoided since the acidity is not concentrated in any one region.

Figure 7.5

Scheme for offsetting the greenhouse warming effect of carbon dioxide by forming a semipermanent layer of stratospheric sulfate aerosols. According to this scheme, sulfur would be converted into COS (carbonyl sulfide) before being emitted into the atmosphere. Because COS is more stable in the atmosphere than is sulfur, over time it would be transported to the stratosphere where it would be photochemically transformed into sulfuric acid that would condense into aerosols that would modify the Earth's radiation balance and alter climate.

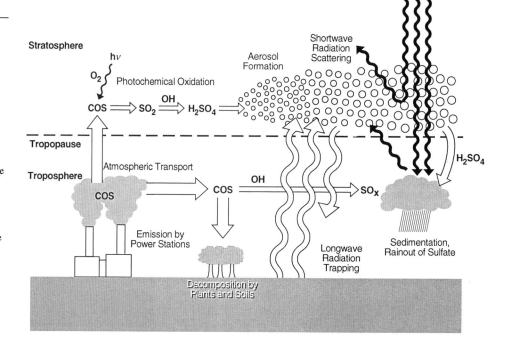

Up to 80 percent of the carbonyl sulfide emitted in the troposphere is destroyed there. The rest is transported into the stratosphere where it is transformed into stratospheric **sulfate aerosol particles.** The observed concentration of COS falls off with height above the troposphere because of photochemical decomposition at high altitudes. The background stratospheric aerosol layer is in large part a consequence of the sulfur liberated from naturally occurring COS. A fraction of the excess COS generated from fossil fuels would therefore enhance the normal stratospheric aerosol layer.

Nearly 50 million metric tons of sulfur could be converted into COS annually in the United States and Europe (equivalent to about 100 million metric tons of COS). Thus, conversion of only 20 percent of this COS into stratospheric sulfate aerosols would be equivalent to the injection of roughly 20 million metric tons of sulfur dioxide into the stratosphere, equivalent to one major volcanic eruption occurring every year. Because this artificial sulfur injection could be maintained over many years, an equilibrium climate cooling of several degrees Celsius would be expected.

The advantages of the COS scheme can be summarized as follows:

1. The solution already lies in the fossil fuels that are causing the problem; the three percent sulfur content of the fuels provides the source of reflective aerosols for mitigating greenhouse warming.
2. The cost of converting sulfur to COS would be low compared to the cost of drastically reducing CO_2 emissions from fossil fuels.
3. The technology for converting SO_2 to COS is quite simple, involving basic thermodynamics and catalytic chemistry. The world economy would not be significantly disrupted during the changeover to COS emission.
4. The excess COS would be widely dispersed and diluted by winds around the planet, eliminating most of the local and regional pollution effects of fossil fuel combustion, including sulfate haze and acid rain.
5. The stratospheric aerosol layer would be formed in a natural way, without the need for aircraft flights or other forms of mass intervention; the cooling mechanism would be similar to that occurring following volcanic eruptions.
6. The thickness of the enhanced aerosol layer, and its duration, could be closely controlled by regulating the rate of COS emission.
7. If the COS emission were halted for any reason, the atmosphere would return to its initial state within a few years (because the atmospheric lifetime of COS is around one year).

There seem to be many advantages to this concept. How could it fail? Unfortunately, it has enormous disadvantages. For one thing, carbonyl sulfide, like most sulfides, is highly toxic. Atmospheric concentrations of COS would have to be enhanced by a factor of 100 to approach 0.1 parts per million by volume. Such high concentrations of COS would pose a worldwide health hazard everywhere and at all times. Near the COS emission sites, the concentrations could be considerably larger, and therefore more deadly.

Moreover, COS, when it decomposes, corrodes metals and causes stomatal damage in plants. Adding insult to injury, sulfides like COS have a powerful odor. For example, the smell of rotten eggs is due to a related sulfide, hydrogen sulfide (H_2S). Any system designed to produce COS would generate an overwhelming stink over a large region.

Another problem with massive COS emissions, and indeed all solutions that propose a stratospheric sulfate aerosol layer thick enough to mitigate climate warming, is that the resulting aerosols would cause the skies overhead to be milky white in color. No more blue skies, ever, anywhere; just an oppressive global pall of haze. (Of course, some may consider that to be only a minor aesthetic issue.)

Stratospheric sulfate aerosols have also been implicated in global ozone depletion. Such particles have been shown to cause ozone destruction, much like the polar stratospheric clouds that are responsible for the ozone "hole" (discussed below). Following the Mt. Pinatubo eruption in 1991, the total abundance of stratospheric ozone worldwide declined by up to 10 percent between 1992 and 1993. That depletion healed as the volcanic aerosols were removed from the stratosphere. In the COS emission scenario, however, the aerosol layer is semipermanent, being renewed continuously for a century or longer. Thus, although the greenhouse effect would be neutralized, the ozone layer would be threatened. The sulfate aerosols themselves would scatter and block some of the dangerous ultraviolet radiation leaking through the depleted ozone layer, but this effect would be insufficient to prevent a net increase of UV-B radiation at the ground.

The three-percent-sulfur "solution" illustrates a common outcome encountered in dealing with environmental problems: Even straightforward technologies can generate nasty side effects. Specialists with quick answers usually turn out to be charlatans.

Fourth of July

Think of the Fourth of July celebration last year: fireworks, smoke and flares, rockets bursting in air, a cool summer night awash in the glare. If one group of technicians has their way, we could be celebrating the Fourth of July all year long on a worldwide scale. They propose to mitigate the greenhouse warming effects of carbon dioxide by filling the stratosphere not with sulfates, but with dust. The method of delivering the dust to the stratosphere, however, is rather unique. It would be lofted using 16-inch artillery shells!

The U.S. Navy has a number of large ships equipped with very large guns that can fire enormous shells at targets miles away. Most of the time, the guns are silent. Why not, it is argued, use those guns to save the environment? Instead of high explosives, the shells can be filled with dust and a small explosive charge, and lofted as high as 15 kilometers, into the lower stratosphere. Like microscopic volcanic eruptions, each shell would add a little dust to the stratosphere. Eventually, a dense layer could be built up and maintained by continuous bombardment. We have the artillery. We have the dust. Why not?

To be effective, perhaps 30 million metric tons of dust would need to be injected into the stratosphere every year. Mineral dust would generally be less effective than sulfate aerosols in creating climatic change, because it is not easy to produce and widely disperse mineral grains with sizes much smaller than one micrometer. Small grains tend to stick together, as in a powder, and are difficult to separate. But larger-size mineral dust grains have two other disadvantages. First, the grains fall more rapidly to lower altitudes, where they are readily removed from the atmosphere. Second, larger particles are less efficient, for a fixed total mass, at increasing the albedo. Hence, to produce the same increase in albedo, considerably more mineral dust would have to be injected into the stratosphere than the amount of sulfate required in the previous scheme.

Suppose that each artillery shell could carry and disperse 100 kilograms (about 220 pounds) of fine dust particles. Then we would need to fire roughly 300 million shells each year, about 10 shots every second of every day for the next century. If 1,000 guns could be made ready for the task (worldwide, the number of such guns that currently exist is perhaps a few hundred), they would need to be fired every minute or so forever. The continuous manufacture of shells would be a problem; shrapnel falling from the skies would be a problem; noise would be a problem. However, one might argue, since the guns are ship-mounted, they can be kept at a distance from populated areas, and moved around to generate a more uniform dust layer. As long as the guns were not pointed straight up, any duds would fall harmlessly into the sea. Still, it is clear that the technical and infrastructure problems with this ludicrous scheme are so profound that it is hard to believe anyone could seriously embrace it.

Sunshades, Balloons, and Boogie Boards

If an increase in planetary albedo will fix the climate, there are a variety of other ways to create it. Everyday experience tells us that white objects reflect more light than black objects, and a mirror reflects almost all the light that falls on it. One idea proposes to place huge solar shades in space. These space "parasols" would reduce the amount of sunlight impinging on the Earth. Only a few percent reduction is needed (this is somewhat less than the increase in albedo required to produce the needed temperature compensation). The area on the ground that would have to be shaded is roughly three million square kilometers. Unfortunately, to create this equivalent shading effect on the Earth from space, a sunshade about 100 times larger would be needed—that is, more than twice the cross-sectional area of the Earth! Such a large size is required because it must be placed at a great distance from the Earth, in a gravitationally stable position referred to as a **Lagrangian point.** In this position, the shade casts only a partial shadow on the Earth, similar to a partial eclipse of the sun; thus, the shade must be correspondingly increased in size to produce the same effective reduction in solar radiation. No technology conceivable today can construct an Earth-sized reflector and place it in the appropriate position.

Several researchers are exploring the idea of using balloons with a shiny metallic coating to increase the planet's albedo. Imagine constructing a fleet of such balloons, filling them with helium and letting them loose in the atmosphere. As they drifted around the world, the balloons would act like tiny clouds, reflecting sunlight away from the Earth. We imagine balloons a few meters in diameter, floating in the lower stratosphere. Scientists already fly balloons there, so it sounds easy enough.

How many reflecting balloons would be necessary to compensate for greenhouse warming? The answer is disconcerting. The effective area of shading required is about 4,000,000 km^2 (roughly one percent of the surface area of the Earth). Assume that the cross-sectional area of a typical balloon is about four square meters, corresponding to seven feet in diameter. (Note that a spherical balloon will not reflect to space all of the sunlight hitting it; depending on the time of day, varying amounts of the reflected light will be reflected toward the Earth.) It follows that at least one *trillion* (10^{12}) balloons are required! That is nearly 200 balloons for every human on the planet today.

Balloons, moreover, are not permanent. Even if carefully designed they might, on average, last one year in the upper atmosphere. In other words, one trillion balloons must be launched every year. If every human alive today launched a balloon each morning, the reflective shield could be maintained.

The problems with such a plan are overwhelming, to say the least. What would the sky look like with a trillion balloons floating around? Occasionally balloons would cluster together, accidentally creating a local solar eclipse. The surfaces of the balloons would strongly affect gases in the stratosphere, like ozone. It is likely that major chemical perturbations would occur. Spent balloons falling from the sky at a rate of several billion per day would foul the countryside, waterways, and the oceans. If longer-lived mini-blimps could be designed, say to last 10 years, they would tend to collect at high latitudes, forming massive superballoon clusters. Putting motors on the balloons to push them around would create other problems depending on the motor and energy technology used.

Instead of balloons, at least one cheaper and faster scheme has been proposed: the "boogie board" solution. Styrofoam is a white, reflective material that floats on water. The oceans have a naturally low albedo—less than 0.1. By floating enough styrofoam on the oceans, the planetary albedo could be increased significantly. The required reflective area is similar to, but larger than, the area that must be shaded by balloons—equivalent to a few percent of the cross-sectional area of the Earth. The oceans cover only about two-thirds of the surface area of the planet, and clouds normally blanket half of the ocean area at any time. Accordingly, the fractional area of the oceans that must be covered by styrofoam would need to be close to 10 percent.

Think of the oceans awash in styrofoam. Life in the seas would be devastated. Alternatively, huge styrofoam continents (something like two Antarcticas) could be constructed and moored in the major oceans. We might also imagine that such continents could serve a number of useful purposes, such as providing space for resorts along a vast coastline. Perhaps guests could be entertained by the fireworks from naval guns filling the stratosphere with dust. One vision of such an engineered world of the future is depicted in Figure 7.6.

Fixing the Ozone Shield

The role of chlorofluorocarbons (CFCs) in reducing stratospheric ozone was outlined earlier. CFCs contaminate the current environment, and those already released will persist well into the next century. The **Antarctic ozone "hole"** and serious worldwide ozone depletion are among the identified consequences. Although a landmark international treaty, **the Montreal Protocol,** has been adopted, which proposes to eliminate CFC production before the turn of the century, it will be decades beyond that time before the ozone layer recovers significantly. What if we cannot wait that long? Suppose that ozone depletion accelerates, as now appears to be happening over the Northern Hemisphere? What technological alternatives might we have at our disposal to preserve the ozone layer? Are there any corrective schemes that make sense? Concepts to save global ozone have surfaced as the ozone crisis has deepened. A few of these schemes are described below.

Figure 7.6

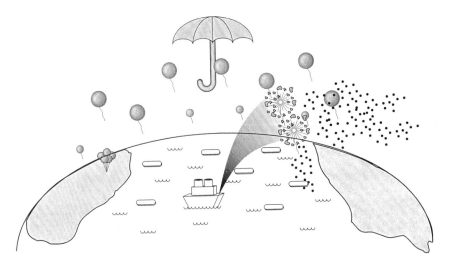

The world as it might look with active global environmental engineering to control climate. The temperature correction schemes shown include lofting artillery shells filled with dust into the stratosphere, releasing a trillion large balloons into the upper atmosphere, and floating huge pads of styrofoam on the world oceans. These schemes, individually or collectively, would seek to enhance the planetary albedo, reflecting light from the planet's surface and thereby reducing the solar effect on the climate. An additional concept (whimsically illustrated by the large umbrella) would place large sun shades in space to decrease solar illumination. The engineering scale and cost of these proposals would be astronomical.

Lasers against CFCs

One problem with CFCs is their long lifetime in the lower atmosphere; there are no known significant sinks for CFCs in the troposphere. Rather, photodecomposition in the stratosphere determines the loss of chlorofluorocarbons. This suggests a rather obvious type of solution—creating an artificial means of depleting CFCs in the lower atmosphere—that has led to a number of proposals.

One idea is to use ultraviolet laser beams to break down CFC molecules. The action would be similar to that which occurs in the stratosphere, where energetic ultraviolet solar radiation does the job. Unfortunately, UV radiation does not travel far in the lower atmosphere, limiting the effective range of the lasers. Moreover, radiation that is capable of destroying CFCs would also attack a wide range of materials. Heavy artificial doses of UV in the troposphere could create all sorts of strange chemical side effects, including choking global smog. What is needed is a magic bullet that destroys only a particular CFC molecule and nothing else, for example, a photon specially tuned to each CFC. However, no such selective photon exists.

However, we might use multiple photons in rapid sequence. The photons would excite a CFC molecule by a series of steps to an energetic state that readily dissociates. Think about climbing a ladder to reach the top of a building from the ground. If the ladder has only one rung, you might have difficulty getting to the top; it is much easier to move up the ladder in small steps, rung by rung. You might also envision spacing the rungs in a particular way that you could negotiate, while a child, say, could not. Analogously, one might imagine raising a molecule level by level from its lowest "ground" state of energy to a higher energy state from which it can fly apart, or photodissociate. The molecule, in this case, is moved rung by rung from the ground by using a series of photons, each of low energy. Every compound has a specific pattern (actually, a number of definite patterns) of photons that can raise it to the photodissociation energy. A sequence of photons that will dissociate one compound will not generally dissociate another. Hence, a highly selective mechanism for photodecomposition is theoretically available—**multiphoton dissociation.**

Multiphoton dissociation does not occur naturally because the intensity of sunlight at each wavelength (or energy) is too low. But a laser is capable of producing an extremely high intensity of light at one specific wavelength. For the greatest effect, the laser beam would be pulsed, and the light emitted in short bursts lasting a millionth of a second or less. Such a laser beam shining on a parcel of air could produce conditions ideal for multiphoton dissociation. The laser must be tuned to the proper wavelength to select the molecule of interest.

For this scheme to work, however, the entire atmosphere must be "laser processed" over a time span short compared to the current atmospheric lifetime of CFCs. A period of 10 years is commonly assumed, although a shorter time span would be preferable. All five billion cubic kilometers of air in the lower atmosphere must be blasted with intense laser radiation. Between 1,000 and 10,000 sophisticated lasers would be required. The total energy requirement has been estimated at several gigawatts to tens of gigawatts.

Unfortunately, such lasers do not actually exist. The costs of building such machines several meters in diameter, generating megawatt beams of radiation, with the specialized optics needed to deflect and guide the rays, would be enormous. Even at that, the proposed laser systems would barely keep up with present emissions of CFCs. To reduce CFC concentrations to negligible amounts in a reasonable time, a project 10 times greater in scale would be called for.

The idea of laser spotlights swinging through the skies all over the world forever is not particularly reassuring. The intensity of the beams would exceed that of sunlight by at least a factor of one thousand. Birds and planes flying through such a beam would be fried. This solution to the CFC problem makes a little sense technically, and makes no sense at all in the global context.

Charging Up the Stratosphere

Another approach to saving ozone while allowing CFC production to continue involves deactivation of chlorine after it is released from CFCs in the stratosphere. Figure 7.7 illustrates such a plan. Neutral chlorine atoms participate in an important **catalytic reaction cycle** that destroys ozone:

$$Cl + O_3 \rightarrow ClO + O_2 \qquad \text{(Eq. 7.3)}$$

However, studies show that chlorine atoms carrying a negative electrical charge will not react strongly with ozone:

$$Cl^- + O_3 \xrightarrow{\text{NO!}} ClO^- + O_2 \qquad \text{(Eq. 7.4)}$$

The scheme, then, is to charge up the chlorine.

A Van de Graaf generator could be flown on an airplane to separate the charges. The positive charge can be transferred to water droplets by bringing them into contact with the positive electrode. If the drops are released from the bottom of the plane, the positive charge will be carried downward into the lower atmosphere. The negative charge remains on the plane, however. As this charge accumulates, electric potentials develop that make it difficult for the positive drops to fall. The proposed solution is to trail a wire behind the plane. Electrons flowing into the wire can leak into the air around the wire as a **coronal discharge.** Such discharges are seen under stormy conditions as glowing light emanating from pointed objects and wires, or dancing from the tops of masts on ships as Saint Elmo's fire. At high voltages, electrons that escape from a surface are accelerated and bang into surrounding air molecules, exciting them to emit visible radiation.

Chlorine is a very "electronegative" element: It likes to acquire and soak up and hold on to free electrons. So the plan outlined in Figure 7.7 to deactivate chlorine might work, right? Wrong! There are so many problems with this scheme it would require an entire chapter to explain all of them coherently. Hence, we will simply sketch a number of the major objections:

1. Chlorine is not the most electronegative substance in the stratosphere. Nitrates and sulfates, which are more abundant, would attract the electrons more strongly than chlorine, preventing the chlorine atoms from accumulating negative charge.

Figure 7.7

A plan to preserve the ozone layer against CFCs (chlorofluoro-carbons) by adding electrical charges to—and thereby deactivating—the chlorine released from CFCs. A fleet of jumbo jets would fly through the stratosphere, trailing wires that emitted a negative electrical charge into the air, and would release water as artificial rain to carry unwanted positive charges to the ground. The charged chlorine atoms would then spiral up to the fair weather electric fields and be accelerated into space.

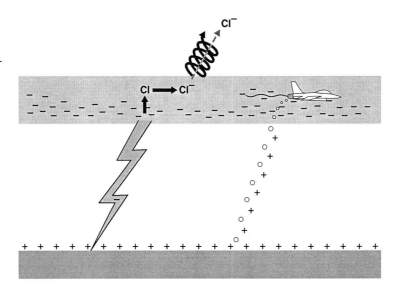

2. Less than 0.01 percent of the chlorine in the stratosphere is in the form of chlorine atoms. The rest is in the form of HCl, ClO, and other compounds. As soon as the free chlorine atoms were charged up, they would immediately be replaced from other chlorine compounds by efficient photochemical processes that operate to maintain a balance between Cl, ClO, and HCl. Indeed, for the scheme to work, all of the chlorine in the stratosphere would eventually need to be electrified. That prospect represents an extremely dangerous conversion of HCl into Cl and ClO, producing conditions similar to those that exist in the ozone hole.

3. Water droplets can carry only a limited electrical charge before they literally explode from Coulomb repulsion. Assuming the optimum charge-carrying capacity of drops, and assuming that only 0.01 percent of the chlorine must be electrified, about 10 billion tons of water would be necessary to separate the required electrical charge. Jumbo jets, each carrying 100 tons, would have to fly 100 million sorties to lift that much water into the stratosphere!

4. The electrical charge separation created—negative charge in the stratosphere and positive charge at the ground—would generate powerful electrostatic fields. Charges would drift in these fields and recombine. The entire system could be discharged within a day by normal atmospheric conduction. Accordingly, the water sorties would have to continue unabated at a rate of 100 million per day essentially forever. The exhaust from these aircraft intrusions would themselves destroy the ozone layer.

5. If the goal of charging up 0.01 percent of the chlorine were achieved—which would have absolutely no ameliorating effect on ozone depletion in any case—the voltage difference between the ground and the stratosphere would be so large that lightning bolts 10 miles long would fly through the atmosphere. Everyone's hair would stand on end, and coronal discharges would be dancing everywhere on the Earth.

The result of attempting to conserve the ozone layer by charging chlorine atoms would actually be the total destruction of ozone and the end of the world as we know it.

Filling the Antarctic Ozone Hole

Measures have also been proposed to prevent ozone depletion on regional scales such as over Antarctica. This ozone reduction, by 50 percent or more, occurs only in austral spring over the Antarctic continent, centered over the South Pole. The motivation for intervention rests on the possibility that the ozone hole may become troublesome during the decades before the Montreal Protocol has had a chance to reduce atmospheric chlorofluorocarbon concentrations significantly. Chlorine, which has been activated over the winter on polar stratospheric clouds, is quickly converted into chlorine, Cl, and chlorine monoxide, ClO, which participate in ozone destruction. Ozone depletion continues until the activated chlorine is redirected into the normally inert chlorine reservoirs, mainly hydrogen chloride, HCl.

The high Cl and ClO concentrations and deep ozone depletions are maintained by the stability of the southern winter **polar vortex**. One thought is to destabilize the vortex. This would abruptly end the conditions favorable for ozone depletion. Alternatively, the Cl concentrations could be suppressed. The critical chlorine catalytic cycles driving ozone destruction involve the reaction of chlorine atoms with ozone. Chlorine monoxide is generated in the process. In the concept considered here, the Cl atom would instead react with light hydrocarbons such as ethane (C_2H_6) and propane (C_3H_8), converting chlorine atoms to HCl. These common hydrocarbons react vigorously with chlorine, and they do not produce any unusual or long-lived byproducts. The reactions of interest are:

$$Cl + C_2H_6 \rightarrow HCl + C_2H_5$$
$$Cl + C_3H_8 \rightarrow HCl + C_3H_7$$

(Eq. 7.5)

The chemical products on the right-hand side of the reactions react further with chlorine; the details are not crucial. The net effect is to convert Cl and ClO into HCl.

The plan would be to add the hydrocarbons to the polar vortex in late winter to inhibit the sudden ozone decrease that appears at first light. The selected hydrocarbons would be spread across the vortex using aircraft (Figure 7.8). These hydrocarbons are stable in polar darkness and would disperse throughout the vortex as winter progressed. When the sun rose in spring to activate chlorine, the hydrocarbons would begin to do their work.

This scheme to plug the ozone hole seems relatively straightforward. Light hydrocarbons like ethane and propane are available, cheap and, aside from the danger of explosion, fairly easy to handle. Propane fuel is already in wide use in the form of liquefied petroleum gas. Calculations suggest that to prevent the ozone hole, about 20,000 tons of liquefied propane or ethane would have to be lifted into the polar stratosphere. A small fleet of specially fitted aircraft could make the flights over a three-week period; 10 planes might be enough. This seems a rather small price to pay to fill the ozone hole, should it become a serious hazard. There is some doubt the plan would work, however.

The ozone hole is quite extensive, covering an area as large as the Antarctic continent and more than 10 kilometers in vertical extent. No aircraft is presently capable of flying over the entire range of altitudes in the vortex where ozone is normally depleted. For the scheme to work, hydrocarbons must be more or less uniformly distributed throughout this vast region. Here, nature is uncooperative. The atmospheric region of interest has relatively weak mixing. It is uncertain whether the hydrocarbon vapors would be spread uniformly enough to "cure" the ozone hole.

Under the unusual atmospheric conditions that characterize the southern winter polar vortex, HCl derived from active chlorine through the action of injected hydrocarbons is not secure. With polar stratospheric clouds present, the HCl can be quickly recycled into active chlorine. In the process, the hydrocarbons would be consumed. Under some not unlikely circumstances, the addition of hydrocarbons might actually exacerbate the ozone depletion problem. Polar stratospheric clouds have been observed to persist at high latitudes far into the spring season. These lingering clouds might cause the injected hydrocarbons to turn against ozone. The result could be prolonged, aggravated depletion of the ozone layer.

It should be emphasized that the depleted polar ozone represents only a few percent of the total global ozone layer. Filling the hole is global environmental engineering on a minor scale compared to some of the other concepts discussed here. Nevertheless, it includes elements of surprise, uncertainty, and awesome scale.

Figure 7.8

Diagram illustrating how the Antarctic ozone hole might be plugged by releasing volatile hydrocarbons (shown here as ethane, C_2H_6) from high-flying airplanes. The hydrocarbons— which react with chlorine atoms (Cl) to form hydrochloric acid (HCl), a compound that does not react directly with ozone (O_3)—are eventually oxidized into carbon dioxide and water in amounts too small to disturb the global stratosphere.

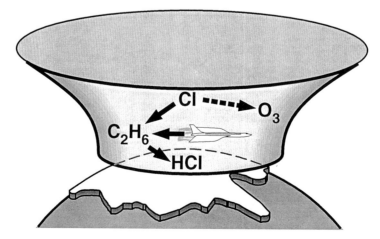

A RATIONAL APPROACH TO ENVIRONMENTAL MANAGEMENT

The application of technology to solve environmental problems is often fraught with pitfalls, especially when the intervention must extend over global distances and span decades to centuries. Some of the examples cited above reflect a lack of basic understanding about the natural environment. Ideas to inject chemicals into the stratosphere or build huge parasols in space seem ludicrous. Extreme caution in environmental intervention is essential if serious mistakes and a significantly reduced quality of life are to be avoided.

To confuse the issue, a cadre of clever pseudotechnicians and political hacks has surfaced, denying that global environmental problems exist in the first place. They find numerous flaws in the generic "disaster" theories of environmental degradation. Thus, we hear about the "holes" in the ozone theory, or the "cooling down" of the greenhouse warming effect. These polemicists are outside of science. By emphasizing the small differences of opinion that often appear between competing researchers and highlighting irrelevant observations that seem to conflict with scientific consensus, they piece together a phony case. Their nonscientific approach is worse than nonsensical; it is dangerous.

Many people, exposed to such arguments highlighted in media hype, have come to believe that technology will always be available as a solution to environmental problems. Technology has been known to perform miracles. Behold the atomic bomb! Behold television! During the late 1950s, Walter Cronkite hosted a television show called "The Twentieth Century." Each week new technological wonders expected to be common in the next few decades were unveiled: limitless nuclear energy; nonpolluting cars magnetically levitated on tracks and automatically piloted; cures for cancer. Few of the wonders promised for this century have actually materialized. Instead, many important technologies have turned against us. Nuclear energy generates a nightmare of radioactive waste. Automobiles foul the air in cities around the world. Chemicals meant for better living fuel an epidemic of cancer. Television devolves into an inane medium of mass marketing.

Education is the key to saving the global environment. High schools and colleges should be places where students learn how the environment works. Every student should gain a basic understanding of the natural world, which ultimately feeds and nurtures our species. Sensitivity to environmental deprivation should be taught universally. Environmental activism could become a normal part of our lives. An appreciation of the benefits of a clean, healthy environment would make the cost of conservation palatable. Every person would have a role in nurturing that lovable global organism called the biosphere.

The **Montreal Protocol** controlling the production of chlorofluorocarbons to save the global ozone shield is an astounding milestone in international environmental law. For the first time in history, the collective population of the Earth has recognized a serious threat to the global environment and acted to fix it. The solution that was finally adopted is *not* a technological cover-up. The source of the ozone depletion problem—chlorofluorocarbons— is being eliminated from all uses. This mandates significant and permanent changes in a major global industry under unprecedented international oversight and regulation. Personal life-styles will be affected by the treaty (which affects the costs for refrigeration, air conditioning, dry cleaning, and so on). In this instance, the changes are likely to be relatively painless. The offending industry is small, and new, ozone-safe compounds are available to replace the older ozone-depleting chlorofluorocarbons. Nevertheless, the transition to a CFC-free world under treaty guidelines over the next decade will require sacrifices, ingenuity, and huge investments of capital. Despite the costs and inconvenience, world leadership is in essential agreement with the decision to proceed.

Too many technologies that have been established by industry and government are dirty and dangerous. The world badly needs clean, safe technologies. That is exactly what a new breed of engineers and scientists, sensitive to the fragility of the environment, are seeking to achieve: non-polluting energy sources; renewable fuels; safe waste disposal; toxin-free foods; pristine air; drinkable water; fumeless transportation. Rather than just "cheapest" and "easiest," another adjective has been added to the design lexicon: "cleanest." Technology is being designed with the environment in mind from the outset, not merely as an afterthought—technology that is sensitive to the ecology of the land and water, and ultimately, technology for people.

This chapter demonstrates that the application of technological schemes to fix environmental problems is generally a mistake. However, technology as a tool in environmental remediation must not be abandoned. In some cases, the environmental risks associated with the small-scale application of a particular technology are well understood. In such cases, small-scale tests could be performed to determine potential safe uses of the technology. Nevertheless, in almost every instance of environmental pollution discussed in this book, the first and most logical approach to remediation is to identify and eliminate the *source* of the pollution. Technology applied to mask or correct undesirable environmental conditions while leaving the cause undiminished should be second, third, fourth—or last—on the list of remedial options.

■

REFERENCES

Broecker, W. 1985. *How to Build a Habitable Planet* (Palisades, NJ: Eldigio Press).

Brown, L., Flavin, C., and Postel, S. 1991. *Saving the Planet.* Worldwatch Environmental Alert Series (New York: W.W. Norton).

Charlson, R., Lovelock, J., Andreae, M., and Warren, S. 1987. Oceanic phytoplankton, atmospheric sulphur, cloud albedo and climate. *Nature* 326: 655–661.

Cicerone, R., Elliot, S., and Turco, R. 1991. Reduced Antarctic ozone depletions in a model with hydrocarbon injections. *Science* 254: 1191–1194.

Cicerone, R., Elliot, S., and Turco, R. 1992. Global environmental engineering. *Nature* 365: 472.

Crutzen, P., and Birks, J. 1982. Twilight at noon: The atmosphere after a nuclear war. *Ambio* 11: 114–125.

Harwell, M., and Hutchinson, T. 1985. *Environmental Consequences of Nuclear War, Volume II, Ecological and Agricultural Effects* (New York: John Wiley and Sons).

LaMarche, V.C., Jr., and Hirschboeck, K.K. 1984. Frost rings in trees as records of major volcanic eruptions. *Nature* 307: 121–126.

McKay, C., Toon, O.B., and Kasting, J. 1991. Making Mars habitable. *Nature* 352: 489–496.

Martin, J., Gordon, R., and Fitzwater, S. 1990. Iron in Antarctic waters. *Nature* 345: 156–158.

Oberg, J. 1981. *New Earths* (Harrisburg, PA: Stackpole Books).

Pittcock, A., Ackerman, T., Crutzen, P., MacCracken, M., Shapiro, C., and Turco, R. 1986. *Environmental Consequences of Nuclear War, Volume I, Physical and Atmospheric Effects* (New York: John Wiley and Sons).

Pollack, J.B., Toon, O.B., Sagan, C., Baldwin, B., and Van Camp, W. 1976. Volcanic Explosions and Climate Change: A Theoretical Assessment. *Journal of Geophysical Research* 81: 1071–1083.

Pollack, J.B., and Sagan, C. 1993. Planetary engineering. In: J. Lewis and M. Matthews (eds.), *Near Earth Resources* (Tucson, AZ: University of Arizona Press).

Sagan, C., and Turco, R. 1991. *A Path Where No Man Thought: Nuclear Winter and the End of the Arms Race* (New York: Random House).

Schneider, S., and Mass, C. 1975. Volcanic dust, sunspots, and temperature trends. *Science* 190: 741–746.

Turco, R., Toon, O.B., Ackerman, T., Pollack, J., and Sagan, C. 1983. Nuclear winter: Global consequences of multiple nuclear explosions. *Science* 222: 1283–1292.

Turco, R., Toon, O.B., Ackerman, T., Pollack, J., and Sagan, C. 1984. The climatic effects of nuclear war. *Scientific American* 251 (2): 33–43.

INDEX